WATER FOR AGRICULTURAL COMMUNITIES

Jide Fatokun

ATHENA PRESS
LONDON

WATER FOR AGRICULTURAL COMMUNITIES
Copyright © Jide Fatokun 2004

All Rights Reserved

No part of this book may be reproduced in any form
by photocopying or by any electronic or mechanical means,
including information storage or retrieval systems,
without permission in writing from both the copyright
owner and the publisher of this book.

ISBN 1 84401 223 9

First Published 2004 by
ATHENA PRESS
Queen's House, 2 Holly Road
Twickenham TW1 4EG
United Kingdom

Printed for Athena Press

WATER FOR AGRICULTURAL COMMUNITIES

ABOUT THE AUTHOR

Jide Fatokun has a B.Sc. (Hons) in Agriculture, M.Sc. in Water Management and a Ph.D. in Water Resources Engineering from the University of Ife, Nigeria (now OAU), Agricultural University of the Netherlands, Wageningen and Trinity University, Dover, Delaware, USA respectively. He was principal of the Bakura College of Agriculture (Irrigation), Ahmadu Bello University, Nigeria; Head of Engineering Department; General Manager and Managing Director of the Ogun-Oshun River Basin Development Authority, Abeokuta, Nigeria; and private Water Resources Development Consultant on IFAD and World Bank-funded projects in many African countries. He is registered by the Council of Registered Engineers of Nigeria (COREN) and is a member of several professional associations and author of seven significant articles in professional journals.

He is married with children and now lives in his home town, Ibadan, Nigeria.

Contents

FOREWORD	xv
PREFACE	xvii
ACKNOWLEDGEMENTS	xix
I. INTRODUCTION	21
1.1 The Place of Water in Agriculture	21
1.2 The Occurrence of Water	23
1.3 Classification of Climates	26
1.4 Hydrology (Agrohydrology)	28
1.5 The Occurrence of Flood	28
1.6 Water Quality	29
II. RAINWATER HARVESTING	31
2.1 Introduction	31
2.2 Rain Harvesting (Detention/Interception)	31
2.3 Roof Catchment	31
2.4 Rock Catchment	33
2.5 Determination of Basic data for Rural Water Supply	37
2.5.1 Water Users' Needs (Water Requirements)	37
2.5.2 Calculation of Run-off Volume from Roofs	38
2.6 Different Types of Tank Used for Rain Water Harvesting	39
2.6.1 Above-ground Tanks	39
2.6.2 Underground Tanks	39
2.6.3 Calculation of Volumes of Water Tanks	40
2.6.3.1 Cylindrical Tanks	40
2.6.3.2 Hemispherical Tanks	40
2.7 Design, Quantities and Construction of Storage Tanks	41
2.7.1A Cylindrical Ferro-cement Tanks	41
2.7.1B Standard Design of Cylindrical Ground Tanks for Capacities in Excess of 30 m^3	43
2.7.2 Hemispherical Underground Ferro-Cement Tank, Extended above Ground Level with a Cylindrical Section	45

2.8 Procedure for Constructing Ferro-cement Tanks	47
2.8.1 Notes	47
2.8.2 Raised Cylindrical Tanks	48
2.8.2.1 Site Preparation and Positioning the Tank	48
2.8.2.2 Preparing the B.R.C. Weld Mesh Frame	48
2.8.2.3 Concreting the Foundation Must be Complete in One Day	49
2.8.2.4 Lining the Tank with the First Coat of Plaster	50
2.8.2.5 Lining the Tank with the Second Coat of Plaster	51
2.8.2.6 Curing the Lining and Coating the Tank	51
2.8.2.7 Roofing	52
2.8.3 Extended Hemispherical Ground Tanks (Figures 2.6 & 2.11)	53
2.8.3.1 Site Preparation and Positioning the Tank	53
2.8.3.2 Excavating the Tank	54
2.8.3.3 Building Instructions	54
2.8.3.4 Roofing	58
2.8.3.5 General Quality Control	60
2.8.3.6 Cleaning and Maintenance	61
III. GROUNDWATER	62
3.1 Occurrence of Groundwater (subsurface water)	62
3.2 Abstraction of Groundwater	64
3.2.1 Exploiting Groundwater Resources	64
3.2.2 Spring Protection	64
3.2.3 Methods of Collecting Water from Springs	68
3.2.3.1 Gravity Springs	68
3.2.3.2 Artesian Springs	68
3.2.4 Wells	71
3.2.4.1 Dug Wells	71
3.2.4.2 Construction of Dug Wells	73
3.2.4.3 Location of Dug Wells	78
3.2.4.4 Improvement of Existing Wells	79
3.2.5 Boreholes	82
3.3 Groundwater Dams	84
3.3.1 Introduction	84
3.3.2 Advantages of Groundwater Dams	85
3.3.3 Factors Which Favour Groundwater Dams	86
3.3.4 User Aspects of Groundwater Dams	89

3.3.5 Economic Factors	91
3.3.6 Cost of Water for Different Materials	91
3.3.7 Planning and Investigating Methods	93
3.3.8 Investigation Methods	93
3.3.9 Design and Construction of Groundwater Dams	94
3.3.10 Water Extraction from Groundwater Dams	97
3.3.11 Dam Construction	98
3.4 Utilisation of High Groundwater Table (Dry Farming)	102
IV. SURFACE WATER	**103**
4.1 Introduction	103
4.2 Utilisation of Wetlands	103
4.3 Diversion/Pumping from Rivers and Lakes	103
4.4 Lake Intakes	107
4.4.1 Introduction	107
4.4.2 Typical Intake Constructions	107
4.5 Dams and Valley Tanks	108
4.5.1 Introduction	108
4.5.2 Ponds	108
4.5.3 Valley Tanks and Dugouts	109
4.5.4 Dams	109
V. REQUIRED STUDIES AND INVESTIGATIONS FOR DAMS AND VALLEY TANKS	**110**
5.1 Identification of the Need for Impounding Water	110
5.2 Studies, Design and Construction of Dams	110
5.3 Site Selection	111
5.3.1 Technical Criteria	111
5.3.2 Non-technical Criteria	112
5.3.3 Field Survey and Investigations	112
5.4 Hydrological Studies	113
5.4.1 Rainfall/Run-off Relationship	113
5.4.2 Description of the Project Area	113
5.4.3 Assessment of Surface Water Potentials	114
5.5 Determination of Scheme/Project Water Requirements and Storage Capacity	118
5.6 Determination of Water Levels and Dam Height	129

5.7 Soil and Geotechnical Investigations	136
VI. DESIGN OF VALLEY TANKS AND DAMS	143
6.1 Design of Valley Tanks	143
6.2 Design of Dams	145
6.2.1 Introduction	146
6.2.2 Plan of Dam and Appurtenant Structures	146
6.2.3 Embankment Type	149
6.2.4 Foundation Treatment	150
6.2.5 Seepage Control	150
6.2.6 Seepage Control in Different Foundation Types	151
6.2.6.1 Introduction	151
6.2.6.2 General	151
6.2.7 Foundation Treatment Methods	151
6.2.7.1 Rock Foundations	151
6.2.7.2 Sand and Gravel Foundations	152
6.2.7.3 Silt and Clay Foundations (Fine-grained Soils)	153
6.2.8 Selection of Embankment Slopes	155
6.2.9 Freeboard	156
6.2.10 Slope Protection and Surface Drainage	157
6.2.11 Spillway Design Discharge	158
6.2.12 Abatement of Sedimentation	161
6.2.13 Drainage and Outlet Structures	161
6.2.14 The Use of PVC Pipes	168
6.2.15 Environmental Protection	170
6.3 Dam Rehabilitation	171
6.3.1 Introduction	171
6.3.2 Kishangura Dam – A Case Study	171
6.3.3 Desilting versus Embankment Raising	171
6.3.4 General Comment	173
VII. BID DOCUMENT, BIDDING, CONTRACTING AND CONSTRUCTION	176
7.1 Calculation of Quantities	176
7.1.1 Introduction	176
7.1.2 Embankment Quantities	176
7.1.3 Spillway Quantities	179
7.1.4 Lateritic Material Capping on Embankment Crest	179
7.1.5 Removal of Old Embankment	179

7.1.6 Opening Up Access Road	180
7.1.7 Area to Clear	180
7.1.8 Area to Grub	180
7.1.9 Area to Desilt	180
7.1.10 Fence Perimeter	180
7.1.11 Area to Grass	180
7.1.12 Water Delivery System	180
7.1.13 Others (see detailed bills)	180
7.1.14 Quantities for Dam Rehabilitation	180
7.1.15 Quantities for Valley Tanks	181
7.2 Bill of Quantities	181
7.3 Engineer's Estimate	181
7.4 Specifications	182
7.5 Diversion During Construction	182
7.5.1 Introduction	182
7.5.2 Diversion/Care of River for Small Earth Dams	183
7.6 Preparation of Bidding (Tender) Document	184
7.7 Bidding, Bid Analysis and Evaluation and Contractor Selection	185
7.8 Contract Award	185
7.8.1 Construction Contract	185
7.8.2 Supervision Contract	185
7.8.3 Defects Liability Period Services	188
7.9 Construction and Post-Construction Services	189
7.9.1 Introduction	189
7.9.2 Construction Procedure	190
7.9.3 Defects Liability (Maintenance) Period	190
VIII. OPERATION AND MAINTENANCE, MONITORING AND EVALUATION	197
8.1 Introduction	197
8.2 Operation	198
8.3 Maintenance	200
8.4 Monitoring and Evaluation	207
8.4.1 Monitoring	207
8.4.2 Evaluation	207

IX. WATER RESOURCES DEVELOPMENT AND ENVIRONMENT MANAGEMENT — 211

- 9.1 Introduction — 211
- 9.2 Environmental Impact Study/Assessment (EIA) — 211
- 9.3 Suggested Steps to Enhance the Attainment of Low Negative Impacts — 212
- 9.4 Effects of Design and Operational Criteria — 213
  - 9.4.1 Design Considerations — 213
  - 9.4.2 Recreation — 213
  - 9.4.3 Reservoir Operation — 213
- 9.5 Watershed or Drainage Basin — 213
- 9.6 Environmental Integration of Land and Water in a River Basin (Watershed) — 214
- 9.7 Causes of Resource Degradation — 215
- 9.8 Criteria for Land-Water Integration in Environment Management — 216
- 9.9 Crucial Action Components to Achieve Integrated Land-Water Conservation and Management — 218
- 9.10 Options for Improvement in Land Productivity — 219
- 9.11 Sedimentation in Reservoirs — 220
  - 9.11.1 The Utility of a Reservoir — 220
  - 9.11.2 Recent Research Findings in Sedimentation — 221
  - 9.11.3 Mitigating Measures — 221
- 9.12 Involuntary Resettlement in Water Resources Development — 222
  - 9.12.1 Introduction — 222
  - 9.12.2 The Objective of Resettlement Programmes — 223
  - 9.12.3 Issues to Consider in Resettlement Planning and Implementation — 223
  - 9.12.4 Resettlement Planning — 224

X. ECONOMICS OF WATER RESOURCES DEVELOPMENT — 225

- 10.1 The Engineer's Problem and the Need for Management — 225
- 10.2 The Engineer's Attempt — 225
  - 10.2.1 The Resources to be Managed — 225
  - 10.2.2 Modifications — 225
- 10.3 The Approach to Engineering Planning — 226
- 10.4 Planning for Water Resources Development — 226
  - 10.4.1 Planning Concept, Involvement and Methodology — 226

10.4.2 Levels of Planning	227
10.4.3 Phases of Planning	227
10.4.3.1 Reconnaissance Study	227
10.4.3.2 Feasibility Study	227
10.5 Preparation for Planning	228
10.6 Planning Objectives	228
10.7 Projections for Planning	229
10.8 Project Formulation	230
10.9 Project Evaluation	230
10.10 Environmental Considerations in Planning: Environmental Impact Study (E.I.S.)	231
10.11 Some Common Pitfalls in Project Planning	231
10.12 Multi-purpose Projects	232
10.12.1 Problems of Multi-purpose Planning	232
10.12.2 Functional Requirement of Multi-purpose Projects	233
10.13 Engineering Economy in Water Resources Planning	234
10.13.1 Introduction	234
10.13.2 Social Importance of Economy in Water Resources Engineering Design	236
10.13.3 Steps in Engineering Economy Study	236
10.13.4 Interest Rates and the Choice of Project Alternatives	239
10.14 Estimated Lives of Hydraulic Structures	240
10.15 Taxes and Other Investment Charges (Capital Recovery Costs)	240
10.16 Probability Concept in Planning	240
10.17 How Hydraulic Structures Should be Designed	241
10.18 Economy Study for Private Versus Public Works	245
10.19 Benefit:Cost Analysis of Alternative Public Works Projects	245
10.20 Capital Budgeting	249
10.21 Cost Allocation for Multi-purpose Projects	250
REFERENCES	255
APPENDIX I: DEFINITIONS	259
APPENDIX II: DAM COSTS AS AFFECTED BY CHOICE OF EMBANKMENT SLOPES, USING KENWA DAM AS AN EXAMPLE	263
APPENDIX III: DESILTING VERSUS EMBANKMENT RAISING IN DAM REHABILITATION	276

APPENDIX IV: KENWA DAM CALCULATIONS	291
APPENDIX V: SAMPLE SPECIFICATIONS AND BILLS OF QUANTITIES	297
A. Sample Specifications	297
1. Introduction	297
2. Definition of Works	297
2.1 Rehabilitation	297
2.2 Valley Tanks	297
2.3 Construction of New Dams	297
2.4 Construction of New Valley Tanks	298
3. The Embankment	298
3.1 The Foundation	298
3.2 Removal of Topsoil	298
3.3 The Clay Core or Water Barrier Portion	298
3.4 Embankment Breaching (for rehabilitation only)	298
3.5 Embankment construction (Core and Shell)	298
3.6 The Dam Wall (or Shell)	299
3.7 Embankment Slope and Crest Protection	299
3.8 Scour/Drainage and Outlet Works	299
3.9 Back-filling	300
3.10 Stilling Basins	300
4. The Spillway Freeboard	300
5. The Environmental Protection Works	300
5.1 Fencing	300
5.2 Grassing	300
6. As-built Drawings	301
7. Specifications	301
B. Bills of Quantities	301
APPENDIX VI: ABRIDGED BIDDING DOCUMENT BASED ON FIDIC FORMAT	311
APPENDIX VII: BID ANALYSIS AND EVALUATION	326
APPENDIX VIII: SAMPLE PAYMENT CERTIFICATES	340

FOREWORD

The importance of water in our daily activities cannot be quantified. Water is essential for human consumption, all aspects of agricultural activities as well as agro-based industries for the development of communities and the sustenance of our civilisation. Water is therefore *life*.

The book, *Water for Agricultural Communities*, is well-packed in ten chapters. It is an invaluable addition to our knowledge of water resources for agriculture. In the introduction, the need for water in agriculture is clearly spelt out.

Chapters II, III, and IV deal in detail with the different available sources of water. It starts with rainwater harvesting, which is of particular interest to areas which receive excess water during the rainy season, but later experience deficit water balance during the dry season. Rain harvesting is therefore an essential farm/village level device for saving excess water during the rainy season for use during dry periods, mainly for domestic and limited agricultural purposes.

Many people, especially in developing countries, still lift groundwater by digging wells (shallow and deep). Groundwater exploitation by means of wells, spring protection, borehole drilling and the use of groundwater dams is copiously covered in chapter III.

Chapter IV covers extensively the harnessing of surface water resources through the utilisation of available wetlands, ponds, valley tanks and dugouts; the diversion/pumping of water from rivers and lakes, as well as the more involved construction of small, medium and even large multipurpose dams. This is followed by the design of dams and appurtenant structures, preparation of bid (tender) documents, submission of tenders, their analysis and evaluation, and finally, contractor selection. I consider this chapter most interesting as it is designed to equip water resources practitioners with adequate knowledge and appropriate steps to take to ensure the successful implementation of dam construction contracts.

After contract award, procedures for effective supervision of dam construction are laid out. It also details the defects liability activities and procedures. This approach will ensure the protection of the interests of both the contractor and his employer (client). Chapter VIII presents the procedures for operation and maintenance (O&M), monitoring and evaluation (M&E) of the performance of the constructed facility; this is a very important aspect of project implementation, which receives due attention in the book.

A major drawback in many water resources projects is their failure to take into consideration their impact(s) on the environment. Dr Fatokun had avoided this pitfall by bringing into focus all that needs to be done to ensure a safe environment for the agricultural community. Equally important are the economics of water resources development, which also receive generous attention in the closing chapter.

In this well-written, easy-to-follow and fascinating book, armed with his vast field experience in many arid and semi-arid African countries, Dr Jide Fatokun had given a detailed account of and the step-by-step procedures for harnessing the different sources of water for agricultural communities, the beneficiaries being farmers and their families, farm workers and rural people for their domestic use, crop production through irrigation, livestock watering, fish farming, as well as agro-based industries. Because of the multifunctional nature of the subject matter so articulately presented in the book, I consider it an indispensable reading to agricultural and water resources practitioners, students in higher institutions of learning, private and public servants, as well as the general reader.

Dr Lekan Are, OON
Ibadan, Nigeria
April 2004

PREFACE

The publication of this book is a milestone in water resources development, specifically intended to meet the requirements of arid and semi-arid regions. Water resources development being a multidisciplinary subject, the book brings under one 'roof' the whole spectrum of the interrelated subjects and disciplines relevant to the planning, implementation, operation and maintenance, as well as monitoring and evaluation of a typical water resources development project.

The sciences and practices of water resources and agriculture are extensive and unlimited in scope and scale; it is the focus of any particular development that determines the scope and scale of coverage. This handbook is aimed at meeting the needs of small to medium agricultural enterprises, the water needs of which can be satisfied by constructing small earth dams and smaller structures. More specifically, it covers small dams, valley tanks, groundwater dams, ponds, roof and rock catchment and direct pumping from lakes and perennial rivers and streams, as well as spring protection and wells – shallow and deep. These are the areas in which agricultural communities form the core of beneficiaries. The focus of water provision from different sources is therefore on crop production, livestock watering and human domestic purposes; these are consumptive uses. Non-consumptive uses like fish farming and recreation can be seen as additional benefits.

The book covers most subject matter areas usually considered in water resources development, from initial briefing by the would-be developer (government, private entrepreneurs etc.) to reconnaissance and subsequent surveys and investigations, detailed designs, preparation of bills of quantities, tender preparation/tendering, tender analysis and evaluation, contract documents, construction, supervision of construction, as well as operation and maintenance, monitoring and evaluation, including environmental impact/studies of water resources development, general water resources and watershed management in existing and/proposed dams/reservoirs. It ends with the economics of water resources development, including multi-purpose projects.

Based on my experience in this field, I have tried to address the needs of different groups and individuals:

- Practitioners of water resources development for all purposes, i.e. consulting agricultural/civil/water resources engineers and contractors
- Students of soil and water conservation/engineering in universities and polytechnics

- Civil servants involved in reading/vetting consultants' reports and
- The general reader.

It is hoped that these and other readers will find the book useful for their different uses and in their different callings.

Jide Fatokun
Ibadan
September 2003

ACKNOWLEDGEMENTS

I am gratefully indebted to many people who helped me in one way or another to make the publication of this book possible, they are certainly too many to mention individually. I must however make specific mention of my wife, Folu Nazaria, and our dear children for their moral and financial support and understanding when I had to be away for long periods in pursuit of this goal; my erstwhile friend, Sunday Ogungbire, for his moral and financial support; Dr Yinka Onabolu for access to certain obscure reference materials and my friend and classmate, Mrs Dupe Awoyomi-Kolade, former Director of Planning, Research and Statistics at the Oyo State Ministry of Agriculture, Natural Resources and Rural Development (Ibadan, Nigeria), for her useful comments and suggestions on chapters VIII through to X; I thank Dr Lekan Are, Officer of the Order of the Niger (OON), my former boss and mentor, for the encouragement he gave me to exploit my writing potential and Archbishop Ayo Ladigbolu of Methodist Church Nigeria for his all-time moral and spiritual support. I thank the copyright holders who allowed me to use materials from their work, and finally, the Almighty God for His Grace so bountifully showered on me, to Him be the glory.

If there is any omission of copyright owners/holders, especially those that could not be traced, I take this opportunity to apologise to any author or institution whose right may have been unwittingly infringed upon. Such infringement will be rectified in subsequent editions.

Jide Fatokun

Chapter I
INTRODUCTION

1.1 The Place of Water in Agriculture

Water is the most prevalent substance on the earth's surface; it covers about 75% of it. It makes up 85% of man's physical being. Even the so-called 'dry land' is frequently charged with and shaped by water. It is the very stuff of life, the major constituent of plants and animals.

Water is a substance of unique and complex behaviour; it constantly dissolves or releases materials, and is subject to frequent changes of state (solid, liquid or vapour) and of properties, which are affected by temperature, pressure and solutes (viscosity, surface tension, etc.)

Water has played and will continue to play significant roles in the development of man's environment. But unfortunately, this very precious resource is sometimes scarce, sometimes plentiful and always very unevenly distributed in space and time. According to the United Nations (UN), some 40% of the world's population suffers from shortage of potable water, while 80% of the Third World's diseases are caused by contaminated water. Because of its importance and uneven distribution, it has become so internationally significant that one frequently hears people say, 'Water is likely to be the cause of the next world war!'

Why do we need to provide water specifically for agriculture? Provision of water as a subject arises only in dry areas where the water resources potential is not only limited and seasonal but also erratic. Surface water potential is a direct function of rainfall, which is generally very low and unreliable in most dry areas. By the same reasoning, groundwater resources are very low since groundwater also comes primarily from rainfall (see hydrological cycle in 1.2). Besides, most dry areas fall within the basement complex rock area, where groundwater potentials are very limited.

Nomadic and semi-nomadic pasturalists, who keep large herds of cattle and hardly engage in crop farming, generally inhabit these dry areas. The scarcity of water does not encourage crop farmers to settle in such areas, while the stocks of cattle farmers are limited by water and pasture, which in itself is limited by water! This situation encourages overgrazing and the spread of livestock diseases because animals cross state/regional boundaries in search of water and pasture. Overgrazing leads to soil degradation and subsequent erosion.

Even when/where some settled life exists based on crop production/animal husbandry, the erratic (uneven) rainfall distribution and frequent droughts in such places reduce life to mere subsistence because it limits both crop and livestock production (reduced crop growth leads to food scarcity and so on).

Provision of water for agriculture is therefore aimed at the following:

A. CROP PRODUCTION

1. Making crop production possible where rainfall is non-existent, as in deserts, or where it occurs in insufficient quantities and duration to make crop production possible.
2. Increasing crop production by:
   i. Ensuring water availability during the growing season (or crop life). This touches on the all-important subject – Crop/Yield Response to Different Moisture Levels.
   ii. Increasing cultivated hectarage and type of crop
   iii. Making planting stocks available always.

CROP/YIELD RESPONSE TO DIFFERENT MOISTURE LEVELS

(i) In considering this subject, it should be noted that the important factor is *plant-water* deficit and not necessarily soil-moisture deficit.

(ii) It should be understood that all other production (growth/yield) factors are not limiting.

(iii) A distinction is usually made between annuals (e.g. grains) and perennials (e.g. citrus); annuals have moisture-sensitive stages of growth and are therefore more likely to respond to variations in soil moisture available than perennials, which are less sensitive (because of better root development) and so tend to even out the effects of different moisture levels in their response.

(iv) While crop response to different soil moisture levels is a rather complex phenomenon (Jackson, 1989), it can be said in general that plant-water deficit, which ultimately results in low or no yield, derives essentially from soil-moisture status. The supply (provision) of water into the soil in adequate quantities at all stages of crop growth is a sine qua non for achieving optimum crop yield, be it annual or perennial crops.

In the above two cases, IRRIGATION is the subject, with the following advantages:

o Providing security against drought
o Ensuring some income for the small farmer from the last (previous season) crops

- Allowing year-round production/cropping by ensuring adequate moisture supply during the moisture sensitive stages
- Increasing crop yield by 3–4 times compared with rain-fed cropping, also by ensuring adequate supply during the moisture-sensitive stages of crop production. Moisture-sensitive stages are discussed in detail in Jackson J (1989)
- Promoting the development of additional areas where rainfall potential is normally poor.
- Promoting crop diversification.

B. LIVESTOCK

- Providing water for optimal production, all things being equal, i.e. ensuring good grass for grazing and water for drinking.
- Reducing the spread of diseases – preventing diseased animals from crossing to uninfected areas in search of water and grass.
- Encouraging settled life among nomads.

1.2 The Occurrence of Water

The hydrological cycle is the cyclic movement of water from the sea to the atmosphere and from there by precipitation to the earth, where it collects by means of small volumes of run-off into streams/rivers and runs back to the sea. The driving forces are the sun's energy and the earth's gravity.

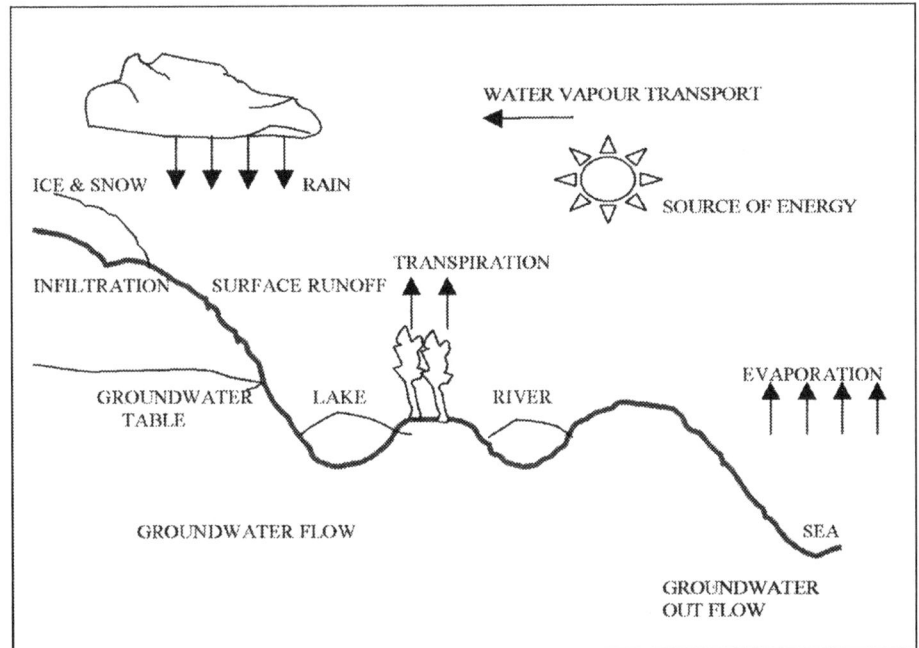

Figure 1.1: The Hydrological Cycle

The cycle can be summarised as follows:

Although the cycle is unending, with no beginning or end, we take the atmosphere as the beginning.

Water from the atmosphere falls to the ground as rain, hail, sleet and snow, or condenses on the ground or on the vegetation. Not all this water adds to surface or groundwater resources, as part of it evaporates and returns directly to the atmosphere. Another part is intercepted by the vegetation or is retained on the ground, wetting the topsoil.

Water accumulating on the ground in pools and marshes is exposed to evaporation.[1] Part of the accumulated water flows as surface run-off towards streams, rivers and lakes. Another portion infiltrates into the ground. This water may flow either at shallow depth underneath the ground to open watercourses, or percolate further downward to reach deeper groundwater strata. Neither the shallow nor the deep groundwater is stagnant; it flows underground in the direction of the downward slope of the groundwater table. Sooner or later the water emerges again at the surface, either in the form of a spring or as a groundwater outflow in a river or lake. From the streams, rivers, lakes, seas and oceans, the water is returned to the atmosphere through evaporation. The whole recycling process then begins again, (Figure 1.1).

By far the greatest part of the water on earth is found in oceans and seas. However, this water is saline. The amount of fresh water is less than 3%, about two-thirds of which is locked in ice caps and glaciers. The fresh water contained underground and in all lakes, rivers, streams, brooks, pools and swamps amounts to less than 1% of the world's water stock.

Most of this liquid fresh water is underground, an estimated 6,000,000 km^3 of groundwater up to 50 metres deep, and a further 2,000,000 km^3 at greater depths. Contrary to popular belief, the amount of fresh water in lakes, rivers and streams is small, about 200,000 km^3 of water. The atmosphere contains only 13,000 km^3 of water. Table 1.1 presents an overview of the average precipitation and evaporation rates for the various continents.

Although this cycle occurs and is visualised to behave as presented above, it is short-circuited at several stages:

- precipitation falls directly into seas, rivers and lakes
- the time the cycle takes varies with the season, e.g. it may seem absent, giving rise to dry weather and the drying of streams – this is the drought phenomenon
- at other times, it seems continuous, giving torrential rains which overtax surface channels, leading to floods
- it is these extremes of drought and flood that are of interest for hydraulic

---

[1]Evaporation occurs from any water surface. Transpiration is loss of water from plants. All plants take water through their roots; it is expelled through transpiration from the leaves.

engineering projects, i.e. dams, spillways and related structures.
- finally, the intensity and frequency of the cycle depends on the geography and climate of a particular place since it operates as a function of solar radiation.

What is presented in Figure 1.1 is therefore a simplification of the processes, which nevertheless gives a good illustration of the natural phenomenon.

TABLE 1.1   PRECIPITATION AND EVAPORATION RATES BY CONTINENT

CONTINENT	PRECIPITATION MM/YEAR	EVAPORATION MM/YEAR	RUN-OFF MM/YEAR
Africa	670	510	160
Asia	610	390	220
Europe	600	360	240
North America	670	400	270
South America	1350	860	490
Australia and New Zealand	470	410	60
Mean values derived after weighting according to area	725	482	243

SOURCE: WMO MONOGRAPH

The quantity and distribution (inventory) of the world's water is summarised in Table 1.2.

TABLE 1.2   QUANTITY AND DISTRIBUTION OF THE WORLD'S WATER

	AREA COVERED (10^6 KM^2)	VOL. IN 10^{12} KM^3	% OF TOTAL VOLUME
Fresh water lakes		125	}
Rivers		1.25	}
Soil Moisture		65	}
Groundwater		8250	} 0.62
Saline lakes and inland seas		105	0.008
Atmospheric water		13	0.001
Polar ice-caps, glaciers and snow	510	29200	2.1
Seas and Oceans	16	1320000	97.25
	360		
TOTAL		1,360,000 OR 1.36×10^{18} M^3	100

SOURCE: ADAPTED FROM E M WILSON & RAUDKIRI

The inventory shows that only about 0.6% of the earth's water is fresh, about half of which is below a depth of 800 m and is therefore practically unavailable. Thus the amount of the earth's water available for man's use one way or another is 4.220625 x 10^{12} or 4 x 10^6 km^3, obtainable from lakes, rivers, soil moisture and part of groundwater. It is from this part of the inventory that water for agriculture derives.

For agricultural purposes, water in the form of overland flow, in streams, rivers and lakes (surface water), and groundwater are the main sources; and for the purpose of this book, the assessment/development of water resources will be limited to surface water and groundwater.

1.3   Classification of Climates

Climatic classification on the basis of monthly, seasonal and annual data, e.g. rainfall – distribution and amount – has been the practice for general purposes. But classification on this basis alone has limitations and is in fact of limited practical application in agricultural production studies. A summary of such classification of dry areas is presented in Figure 1.2, while the characteristics are described below it.

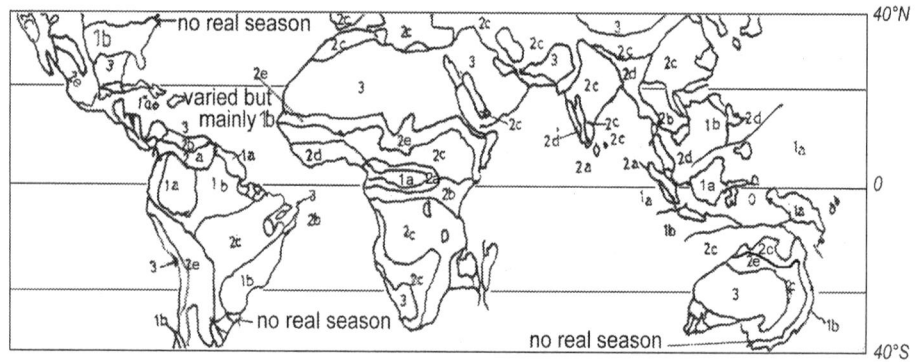

Figure 1.2: Distribution of Seasonal Rainfall Types

(2b) I   Dry tropics   R = 650–1500 mm, 2 short rainy seasons, separated by a pronounced dry season.

(2c) II   Dry tropics   R = 650–1500, one fairly long rainy season, one long dry season.

(2d) III Dry tropics   R > 1500 mm, one exceptionally heavy rainy season and one dry season.

(2e) IV Dry tropics   R = 250–650 mm, one short rainy season and one long dry season.

(3) V   Dry climates   R < 250 mm, little rain at any time, which can be concentrated in one very short 'wet' season of only a few weeks.

But classification on the basis of available soil moisture is more relevant for agricultural purposes. Unfortunately, there does not seem to be such a classification of general applicability at present. It suffices therefore to conclude that all places where rainfall in any month of the growing season is less than 100 mm (R<100 mm), are semi-arid areas, where the provision of water for agricultural purposes by artificial means (man's efforts) is necessary for survival and growth.

Water deficit exists if the water supplied from the soil through precipitation is less than water removed from the soil by the combined actions of evaporation at the soil surface and transpiration from plants growing in the soil (evapotranspiration). This is the situation in months of the growing season, when rainfall is less than 100 mm. Some of the major characteristics of these areas are:

A. DRY TROPICS

I. Annual precipitation is between 650 and 1500 mm. There are two distinct rainy seasons separated by a pronounced dry season (a few months of R<25 mm) and a short drier season. These are mainly in tropical Africa, situated more or less astride the equator, e.g. Uganda.

II. Annual rainfall is between 650 and 1500 mm in one fairly long rainy season (usually 3–5 months, each with R>75 mm) and one long dry season as in northern Nigeria above longitude 10°N, the north and eastern coast of Australia and the bulk of central Africa and Central South America.

III. Annual rainfall is >1500 mm, occurring in one season of exceptionally high rainfall and one long dry season; this is typified by the southern coastal area of West Africa and the monsoon rainfall areas of India and Russia.

IV. Total annual rainfall is between 250 and 650 mm in one short rainy season (3–4 months, each with rainfall >50 mm) and one long dry season. Noticeable areas of this climate are Central and Western parts of Southern Africa, parts of the western coastal areas of the USA and the semi-arid area which borders the southern part of the Sahara Desert.

B. DRY CLIMATES

V. Annual rainfall is less than 250 mm, has little rain at any time, but the rainfall can also be concentrated in a very short 'wet' season, perhaps of only a few weeks.

Rainfall in most of these areas is concentrated within short seasons, especially in Africa and the Monsoon Asia. On the whole, over two-thirds of Africa has over half of its annual rainfall in a period of 3 months as in the semi-arid zone of Northern Nigeria, where 70–90% of the rain occurs in July, August and September.

The water available to plants depends primarily on the soil water potential and hydraulic conductivity, both of which decrease as soil water content decreases.

Detailed discussion of these phenomena is beyond the scope of this book, but in broad terms, the water available to plants is the water between field capacity (FC) and permanent wilting percentage (PWP). While FC is a soil property, PWP is a plant characteristic, occurring when leaves lose their turgor; it depends on plant factors which influence water loss (e.g. transpiration) and water uptake from the soil.

In all the areas described above, the concentration of rainfall within short periods creates a major agricultural problem. Because soil moisture is limited, plants are under stress in the dry season, which happens to be quite long; an important need is thus established to store surplus water (run-off) and conserve it for use in the dry season.

1.4   Hydrology (Agrohydrology)

Hydrology is the science of the occurrence and movement of water over the earth's surface; it deals with the various states – liquid, gaseous and solid. While water in the gaseous form is very important in agriculture (for plant photosynthesis, respiration etc.), that state is only a means to an end, (as in transpiration) and will be considered only in so far as it affects the water required in a project – water requirement. Otherwise, water will be treated only in its liquid state. Agrohydrology (or agricultural hydrology) is hydrology as applied to agriculture, i.e. crop and animal production.

The consideration of hydrology per se not being the subject of this handbook, its consideration will be limited to the assessment of the water resources potential of an area, (catchment) as well as the capacities of storage structures, spillways and related structures. These are the areas of hydrology which the water resources engineer needs to be familiar with. It should be noted however that hydrology is not a precise or exact science; it deals with the probability of the occurrence of extreme events like flood, rainfall of a particular amount and duration, drought etc. Readers who are interested in more detailed hydrology may refer to the texts listed in the References.

1.5   The Occurrence of Flood

When rainfall becomes intense or prolonged or both, the surplus run-off becomes too large for streams and rivers (or smaller drainage channels as the case may be) to contain. They therefore fill and overflow their defined channels and banks. This defined channel overflow, which is called flood, results in water remaining on the surface for prolonged periods and/or flowing rapidly in large volumes. In both cases, enormous damage is done to human activities, the most serious effect of flooding in agriculture being the creation of anaerobic (absence of air) conditions in the root zones of crops and washing away of fertile topsoils in which crops are grown by the receding flood. In flat areas, it is this anaerobic condition in the soil that leads to crop loss.

In agrarian societies, floods are feared because they can destroy crops, cattle and homesteads and so bring famine on communities.

In urban areas, on the other hand, flood causes great damage to property, pollutes water supplies and disrupts sewerage and waste-water disposal systems, as well as endangers lives. Communication systems are also often disrupted. Flood mitigation is outside the scope of this handbook.

1.6 Water Quality

Water quality refers to how suitable the water is for a purpose or purposes. In an agricultural community, water is required essentially for human consumption, livestock watering and crop production, especially irrigation. Irrigation has specific water quality requirements, and while livestock are adapted to poor quality water, human water quality requirements are more stringent.

All sources of water which the farmer comes into contact with derive from rainwater via the hydrological cycle. Although rainwater contains elements dissolved in it as it falls through the atmosphere, it is the purest source for human consumption. But once it comes into contact with surfaces like vegetation, bare soil etc, it picks up other substances, most of which reduce its quality (pollute it) for human consumption.

It is therefore necessary to be aware of the likely health hazards, especially to humans, inherent in each source of water discussed in subsequent chapters.

The World Health Organization (WHO) has developed Guidelines for water quality standards for Human Drinking Water supply. For the sake of completeness, the standards for the major elements usually dissolved in water are given in Table 1.3.

But for a typical agricultural community, general and easily identifiable guidelines will suffice, for example.

1. Good quality water should be colourless, odourless, free of suspended solids, not hard etc.
2. If rainwater is collected from clean roofs and into clean containers, it is the best source of water for drinking purposes.
3. Surface water, i.e. from ponds, streams and similar sources, is the most easily susceptible to pollution, especially by human and livestock faecal materials, agro-chemicals etc. It should therefore be disinfected before drinking.
4. Groundwater from shallow sources (i.e. less than 10 m deep) may also be polluted by faecal matter as from pit latrines and septic tanks.

Therefore, when considering which source of water to develop especially for drinking purposes, the following points should be considered:

1. Rainwater is the first choice if there are corrugated iron sheet or other suitable alternative roofs, it is also usually very near the home.

2. Where springs are available, i.e. water that has been filtered through several soil layers and comes out under certain hydro-geological conditions, it is the second choice for quality purposes, though it could be far from the house; but it also needs to be protected from pollution.
3. Groundwater from depths of 10 m and deeper comes next, i.e. dug wells and boreholes.
4. Where other sources are either not available or costly to exploit, e.g. borehole, then surface water becomes the obvious choice. It will however require some treatment before it can be safely used for drinking and other domestic purposes.

This subject is further touched upon under each source of water discussed in subsequent chapters. For the final word on water quality for human consumption, the engineer should consult the local health/water supply authorities.

TABLE 1.3   GUIDELINES FOR DRINKING WATER QUALITY

WATER QUALITY PARAMETER	MEASURED AS	HIGHEST DESIRABLE LEVEL	MAXIMUM PERMISSIBLE LEVEL
Total dissolved solids (TDS)★	mg/l	500	2000
Turbidity	FTU	5	25
Colour	mg Pt/l	5	50
Iron	mg Fe^+/l	0.1	1.0
Manganese	mg Mn^{++}/l	0.05	0.5
Nitrate	mg No_3^-/l	50	100
Nitrite	mg N/l	1	2
Sulphate	mg So_4^{--}/l	200	400
Fluoride	mg F^-/l	1.0	2.0
Sodium	mg Na^+/l	120	400
Arsenic	mg As^+/l	0.05	0.1
Chromium (hexavalent)	mg Cr^{6+}/l	0.05	0.1
Cyanide (free)	mg CN^-/l	0.1	0.2
Lead	mg Pb/l	0.05	0.10
Mercury	mg Hg/l	0.001	0.005
Cadmium	mg Cd/l	0.0005	0.010

★This includes the major dissolved solids such as SO_4^{--}, Cl^-, HCO_3^-, Ca^{++}, Mg^{++} and Na^+. The levels indicated depend on the climate, the type of food, and the work load of the water user. In some recorded cases, people have lived for months on water having a total dissolved solids content in excess of 5000 mg/l.

Chapter II
RAINWATER HARVESTING

2.1 Introduction

Water for agriculture derives primarily from rainwater harvesting, groundwater and surface water, apart from rain-fed agriculture, which is watered directly by the rain. It also provides water for agro-processing and farm construction purposes. Rainwater harvesting is discussed in this chapter while groundwater and surface water are covered in subsequent chapters.

2.2 Rain Harvesting (Detention/Interception)

This is the interception of rainwater before it reaches the earth's surface and develops into run-off. Because of the limited quantities that can be handled in this manner, water from this source is mainly for human and livestock use. The main methods of harnessing water in this case are:

- roof catchment
- rock catchment (water gathered in this manner serves mainly for irrigation and stock watering, but if it has to be consumed by humans, it has to be given adequate treatment and disinfection.

These methods are aimed at intercepting rainwater and storing it in locations as close as possible to the end users.

2.3 Roof Catchment

This is the collection of rainwater from rooftops – usually corrugated galvanised iron sheets.[2]

Where the conditions allow, an artificially paved catchment ground, connected to a pump and tank, as shown in Figure 2.1B, can also serve where iron sheet roofs are uncommon. This is discussed further under Rock Catchment.

Rainwater which is detained by and runs down the roof, is usually collected in one or more of the following ways:

---

[2] The use of large polythene sheets, nicely sewn together and put on a sloping frame with gutters at the lower end, is a good alternative to corrugated iron sheets, especially in poor rural communities. See Figure 2.1.

- direct catch into household vessels, e.g. buckets, pots, drums
- above-the-ground metal or ferro-cement tanks and
- below-ground brick/stone masonry tanks.

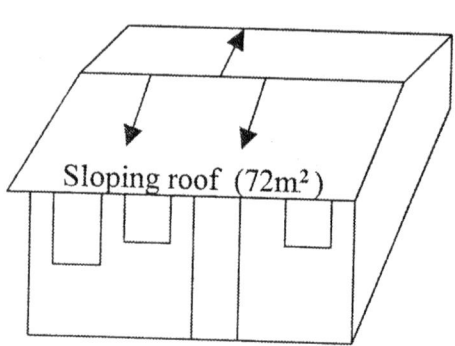

Figure 2.1A: Sloping Roof (72 m^2)

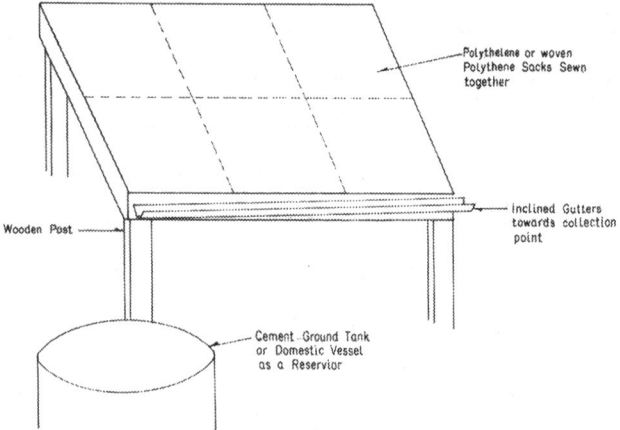

Figure 2.1B: Polythene Sheet/Sewn Polythene Sacks on Sloping Frame Channelled into a Ground Tank

Criteria for siting roof catchment devices

- roofing material should be galvanised iron sheets, to ensure clean and uncontaminated water (sewn polythene sheets are a good alternative in poor communities)

- the entire roof area should not be covered by trees or other canopies – it should be fully exposed to rain drops to prevent contamination and ensure copious catch.
- The roof should be in good condition for the attachment of gutters.
- The roof edge should be at least 2.5 m above the ground to prevent contamination by raindrop splash on the surrounding ground.

2.4 Rock Catchment

Rock catchment is sometimes described as interception of overland flow (run-off), but because the rainwater has no chance of infiltrating into the soil, it is regarded here as rain harvesting. It is possible in rocky areas, where naturally sloping rock surfaces occur and convey rainwater down the slope, where it can be collected in purpose-built reservoirs – generally earth-fill embankments or gravel stone-fill or even holes excavated into the rock itself. The principle of harvesting is the same as for Roof Catchment, varying only in the type and placement of the tank for collection.

There are two types of rock catchment Tanks:

Type 1 is built at the end of naturally occurring exposed rock surfaces, sometimes surrounded with low walls, to channel only water falling on the rock surface into the reservoir, which should be at least 3.5 m deep to minimise evaporation.

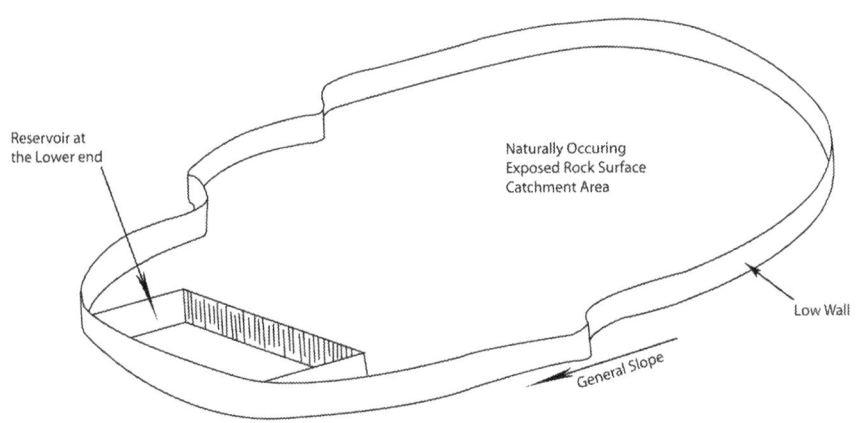

Figure 2.2: Type 1 Rock Catchment
Naturally Occurring Exposed Rock Surface as Catchment Area

Table 2.1 shows the bill of quantities for type 1:

TABLE 2.1   BILL OF QUANTITIES FOR A TYPE 1 ROCK CATCHMENT TANK

S/N	DESCRIPTION	UNIT	QTY	COST (use appropriate local rates and currency)	
				RATE	AMOUNT
1	Clean sharp sand	m^3	2		
2	Cement	bags	10		
3	Standard building bricks	pieces	4600		
4	Corrugated galvanised iron tank	no	1		
5	¾" G.I. pipe	no	5		
6	¾" sockets	no	4		
7	¾" 90° bend socket	no	4		
8	¾" bibcock tap	no	1		
9	Semi-rotary handpump	no	1		
10	Excavation explosive charges (1 box of 25 sticks of dynamite complete with cortex, pot fire, safety fuses and detonators)	box	1		
11	Mason & unskilled labour helper	no	varies		
	TOTAL	-	-		

Type 2 rock tanks are built into the hard firm rock, using the entire rock catchment area, channelling the water through a collector channel, some times equipped with silt traps, into the reservoir. Excavation is done manually with pickaxes, spades, hoes, hammers and buckets – the use of explosives is recommended, but the services of experienced artisans should be employed for this purpose. Where rock fractures occur, they should be sealed with cement slurry to minimise leakage.

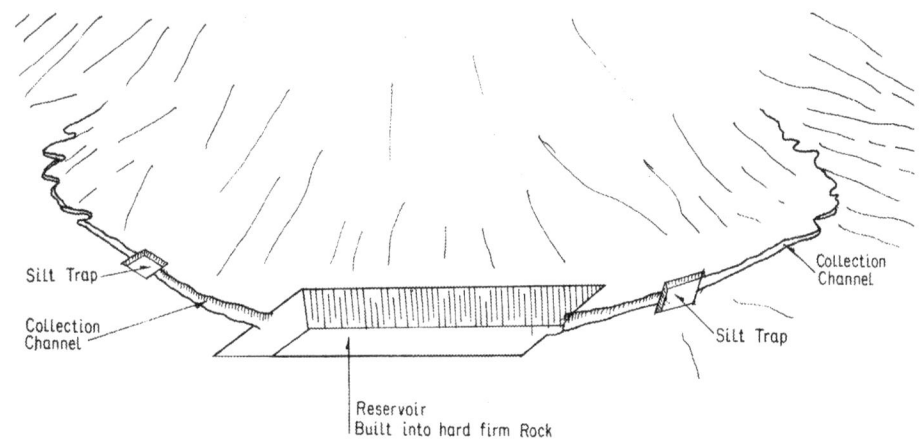

Figure 2.3: Type 2 Rock Catchment –
Uses the Entire Slope, Regardless of cover to collect Rainwater

Table 2.2 shows the bill of quantities for a typical type 2 construction:

TABLE 2.2   BILL OF QUANTITIES FOR A TYPE 2 ROCK CATCHMENT TANK

S/N	DESCRIPTION	UNIT	QTY	COST (USE APPROPRIATE LOCAL RATES AND CURRENCY	
				RATE	AMT
1	Clean sand	m^3	2		
2	Cement	bags	5		
3	Standard building bricks	no	4,600		
4	Corrugated GI tank	no	1		
5	¾" G.I. pipes	no	5		
6	¾" sockets	no	4		
7	¾" 90^0 bend sockets	no	4		
8	¾" bibcock tap	no	1		
9	Semi-rotary handpump	no	1		
10	Excavation explosive (as for Type 1)	no	1		
11	Skilled & unskilled labour	no	varies		
	TOTAL	-	-		

In both cases, the catchment areas range from tens of square metres to hundreds of square metres, with corresponding varying dimensions of storage tank. The water collected can then be used for all kinds of purposes – domestic, livestock watering or even irrigation, depending on the volume collected and the needs of the collector.

Type 3 tanks are artificially paved catchment surface tanks. Because there is no control over the catchment area that contributes rainwater/run-off into the two rock catchment tanks discussed above, the water collected via the two methods often ends up being dirty and polluted, especially with faecal pollutants. To get around this, and in the absence of suitably inclined rock surfaces, artificially-paved catchment surfaces can be constructed. The following is the procedure for constructing one.

1. During a rainfall, identify the general direction of flow of run-off, which gives the general direction of the slope in the area.
2. Select a suitable site, clear and level it and then construct a suitable detention reservoir or tank at the lower end; a reservoir of 225 m^3 with dimensions of L = 15 m, W = 3 m and D = 5 m is shown below.
3. Construct as shown in the figure, using standard building bricks/blocks
4. Line the tank with brick masonry and plaster with sand/cement mortar.
5. Create a paved surface, using bricks or concrete (sand, cement, aggregate mix) or precast pavement slabs, whichever is cheaper.
6. Cover the paved surface with grass, leaves or polythene sheets to allow for setting.
7. Once set, remove the grass and make a channel to lead the water into the channel

A 100 m^2 surface area, collecting water into the 225 m^3 reservoir, filled at the end of the rainy season has been found from experience to provide enough water for 10 families of an average size of 8 per family throughout a 3-month dry season in Uganda (personal experience). See bill of quantities in table 2.3.

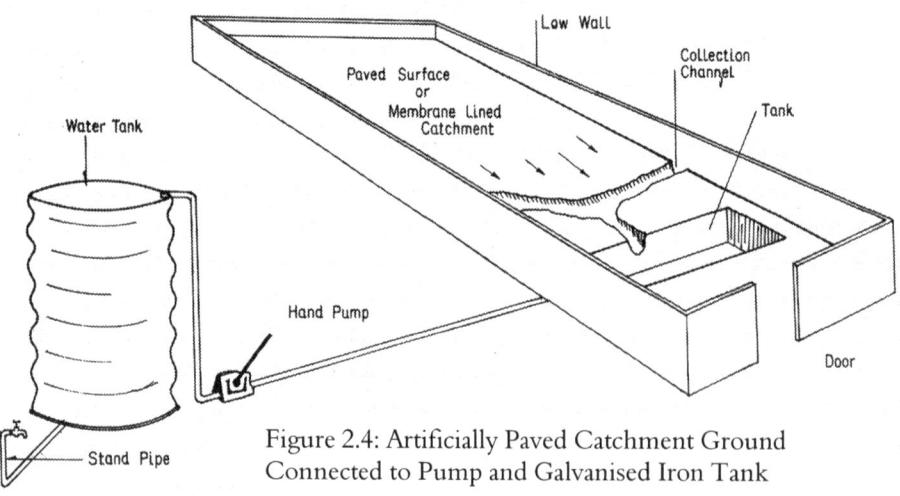

Figure 2.4: Artificially Paved Catchment Ground Connected to Pump and Galvanised Iron Tank

TABLE 2.3 BILL OF QUANTITIES FOR AN ARTIFICIALLY PAVED CATCHMENT GROUND TANK

S/N	DESCRIPTION	UNIT	QTY	COST (use appropriate local rates and currency)	
				RATE	AMT
	Cement	bags	20		
	Building bricks	no	8,550		
	Clean sharp sand	tonnes	3		
	Corrugated G.I. tank	no	1		
	¾" G.I. pipes	no	5		
	¾" 90⁰ bend sockets	no	4		
	¾" bibcock tap	no	1		
	Semi-rotary handpump	no	1		
	Skilled & unskilled labour for excavation & plumbing & related works	no	varies		
	TOTAL	-	-		

2.5 Determination of Basic Data for Rural Water Supply

2.5.1 WATER USERS' NEEDS (WATER REQUIREMENTS)

Determination of water needs requires the following data, among others: average daily consumption per person (per capita daily consumption usually given in litres per capita per day (lcd), this is 20–40 lcd, if the source is near and 10 lcd if the source is far. For planning purposes for rural areas, the figures presented in Table 2.4 are usually adopted:

TABLE 2.4 PER CAPITA DAILY WATER REQUIREMENT FOR RURAL AREAS

Humans	50 lcd	
Grade bull	50*	1.0 L.S.U.
Cow	40	0.8 or 0.7 L.S.U.
Heifer	30	0.6
Calf	10	0.2
Pigs	15	0.3–0.4
Goats/sheep	6.25	0.15
Rural households	20	
Schools	5	

SOURCE: UGANDA NATIONAL STUDY IN SUPPORT OF THE INTER-GOVERNMENTAL NEGOTIATION COMMITTEE ON DROUGHT AND DESERTIFICATION, MAAIF, 1993

---

*Referred to as Livestock Unit (L.S.U.)

These figures are further multiplied by a factor of between 1.1 and 1.3 to cater for incidental population increases and inefficiency of distribution, since it is difficult to accurately estimate future water demand.

Other parameters are:
- Number of days over which the requirement is determined, usually a dry season
- Number of people (and maybe livestock) to be catered for
- Number and type of livestock to be catered for.

For a household of 6 people over a 180-day dry season, the water requirement will be: 180 days x 20 lcd x = 21,600 litres or 21.6 m³. This subject is covered in greater detail is chapter V.

2.5.2   CALCULATION OF RUN-OFF VOLUME FROM ROOFS

The data required are the area of the roof and the annual rainfall in the area for a below average rainfall to ensure reliability of the tank and other structures in years of poor rainfall. 50% of such rainfall is usually taken as the available rainfall. Given an annual rainfall of 620 mm, the available rainfall = 620 x50/100 = 310 mm. Over a roof area of L = 12.85 m and B = 7.65 m, i.e. 98.3 m², the available rainwater in this case = 98.3 m² x 310/1000 m = 30.5 m³ or 30,473 l (see Figure 2.5). This procedure can be applied to any place where the required data are available.

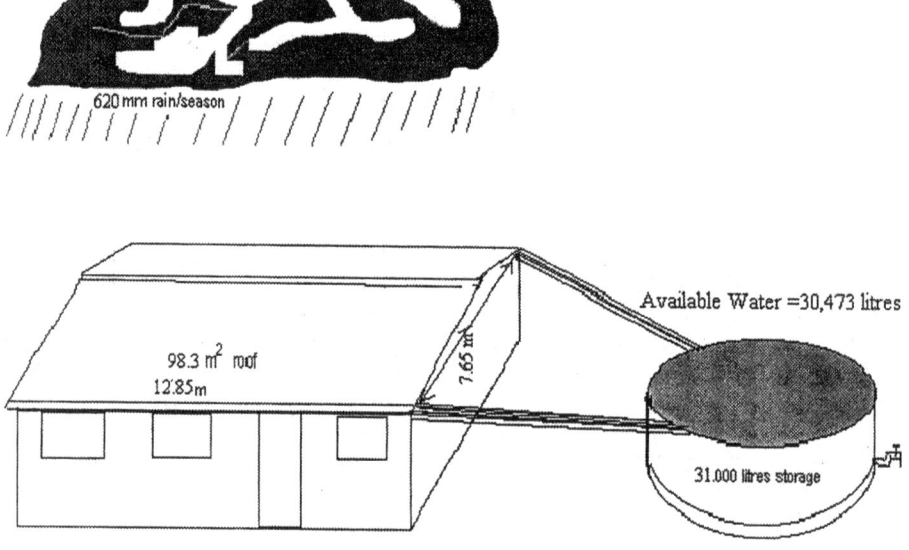

Figure 2.5: Run-off Volume from Roofs

2.6 Different Types of Tank Used for Rainwater Harvesting

2.6.1. ABOVE-GROUND TANKS – METAL, FERRO-CEMENT OR BRICK MASONRY

The arrangement is as shown in Figures 2.4 and 2.5. The method of drawing water from the tank is by means of stand taps or pumps in more affluent communities/homes.

2.6.2 UNDERGROUND TANKS

Square, rectangular, cylindrical or hemispherical tanks of dimensions suitable for the expected storage volumes are dug and lined with bricks/stones and plastered. Concrete slabs, iron sheets or other forms of roofing are constructed over the hole and a manhole provided for water delivery and cleaning. Figure 2.6 is a hemispherical underground tank extended above the ground with a cylindrical wall.

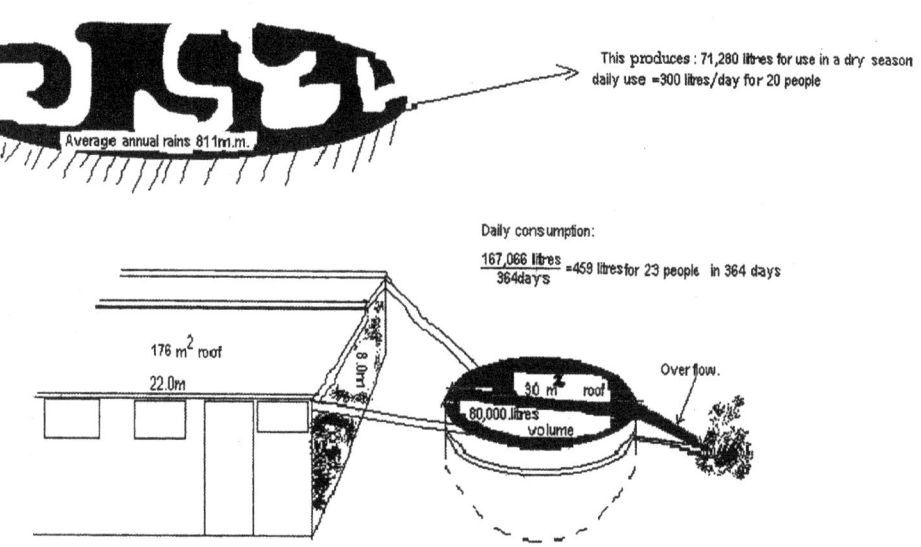

Figure 2.6: Hemispherical Underground Tank with a Cylindrical Above-ground Extension

The common method of drawing water from these tanks is by line and bucket. But pumps – simple hand pumps or semi-rotary – are also common in more affluent societies. The calculation of the volume of water from roofs as well as tank volumes is treated in 2.5.2 and 2.6.3.

2.6.3 CALCULATION OF VOLUMES OF WATER TANKS

2.6.3.1 Cylindrical Tanks
The tank capacity is calculated from the equation: $V = \pi r^2 h$
  Where $\pi = 22/7$
  r = radius of the tank and
  h = height of tank

For a tank of r = 1.87 m and height = 1.9 m
  $V = 22/7 \times (1.87)^2 \times 1.9$
  = 20,881 lit or 21 m³

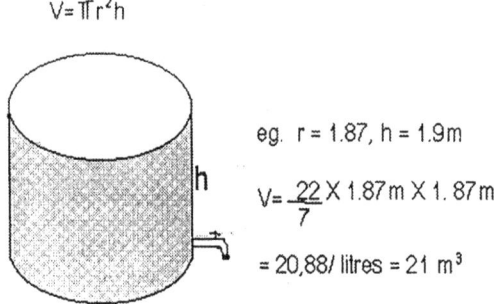

Figure 2.7: Cylindrical Tank

It is common to find 10 m³ capacity tanks on most private and community-based ranch water schemes with the following dimensions:

  Height   2.0 m
  Diameter 2.5 m
  Radius   1.25 m
  Capacity 10 m³ (actual = 9.8 m³)

They are generally constructed of Ferro-cement materials on reinforced frames. More details on design and construction are given in 2.7.

2.6.3.2. Hemispherical Tanks
For a hemispherical tank, there are two parts, the upper cylindrical part and the lower semi-spherical part.

The lower part is a semi-sphere. Volume of a sphere = $\pi d^3/6$, where d = diameter of the sphere. For the tank shown below:

Volume of lower part (V_1):   = $\pi d^3/6 \times 2$
  = 22/7(6.14 x 6.14 x 6.14)/(6 x 2)
  = 60,625 l or 60.625 m³

The upper cylindrical part (V_2): $= \pi r^2 h$
$= 22/7 \times (3.07)^2 \times 0.65$
$= 19,236$ l or 19.236 m^3

Total Volume of Tank $= 60.625 + 19.236$
$= 79.861$ m^3
$= 80$ m^3

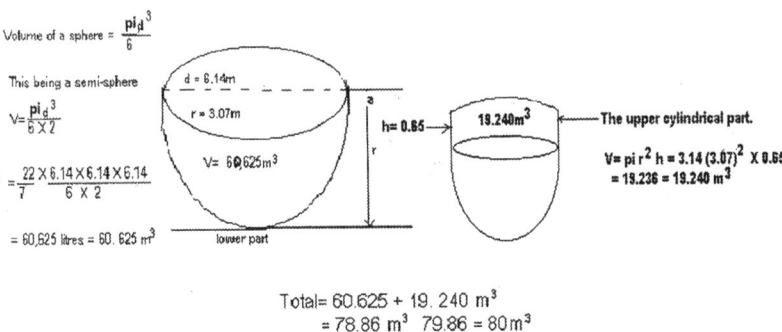

Figure 2.8 Hemispherical Water Tanks

2.7 Design, Quantities and Construction of Storage Tanks

2.7.1A CYLINDRICAL FERRO-CEMENT TANKS

Ferro-cement tanks, roofed with iron sheets or ferro-cement are cheaper than equivalent tanks of bricks, blocks or even galvanised iron, yet they are stronger and more durable, hence, their preference to the conventional tanks is obvious.

The design is the same for the various sizes but a 23 m^3 capacity tank is used as an example. The arrangement in relation to the roof as well as design is represented in Figures 2.9A, B & C.

In case it is not possible to adopt the ferro-cement option, the design of a masonry tank with concrete roof is presented in Figures 2.10A, B and C. The materials estimates are also given for different tank sizes up to 100 m^3 (100,000 litres).

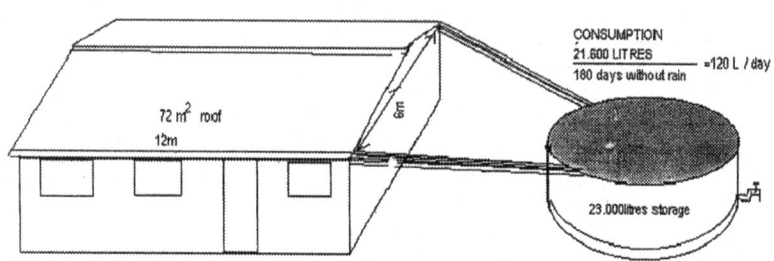

Figure 2.9A: Location and Arrangement of a Raised 23 m³ Cylindrical Tank

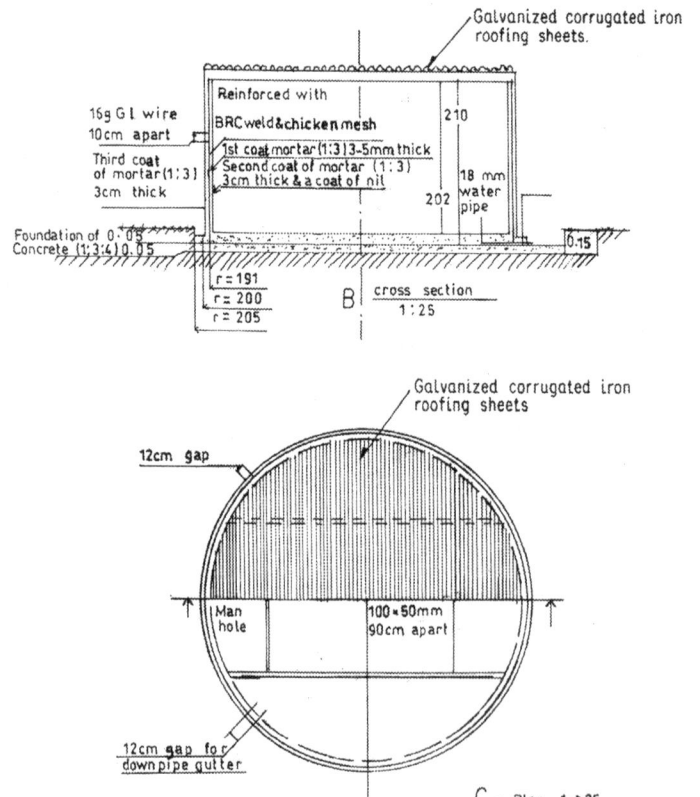

Figure 2.9B & C: Standard Design of a 23 m³ Capacity Cylindrical Ferro-cement Tank

The bill of quantities is given in Table 2.5.

TABLE 2.5 BILLS OF QUANTITIES FOR A 23 M^3 CAPACITY CYLINDRICAL FERRO-CEMENT TANK

S/N	DESCRIPTION	UNIT	QTY	COST (use appropriate local rates & currency)	
				RATE	AMT
1	Cement	tons/bags	1.250–25 bags		
2	B.R.C. mesh no 65 or 66	roll	1		
3	Chicken mesh, 25 mm x 1 m x 30 m	roll	2		
4	Binding wire	kg	10		
5	Draw-off standpipe and tap	set	1–see sketch		
6	Empty polythene bags (sugar/flour)	no	20		
7	Sisal twine	kg	2		
8	Clean sharp sand	tonnes	4-32 w.barrows		
9	Clean ballast or aggregates (5 cm)	tonnes	2–16 w.barrows		
10	Water	tonnes	2–10 drums		
11	Skilled labour	labour-day	24		
12	Unskilled labour (porters)	labour-day	48		
13	Transport	lump sum	to be inserted		
	TOTAL	-	-		

NOTE: Convert tonnes to the more familiar units as follows:
  1 tonne of cement = 20 bags
  1 tonne of sand = 8 wheelbarrows
  1 tonne of stones = 8 wheelbarrows
  1 tonne of water = 5 drums

2.7.1B STANDARD DESIGN OF CYLINDRICAL GROUND TANKS FOR CAPACITIES IN EXCESS OF 30 M^3

The standard practice is to design tanks in this category as in Figures 10A, 10B and 10C. The diameter of such tanks exceed 3.5 m; in practice, it is more economical to support the roof with a G.I. column as shown in Figure 10B.

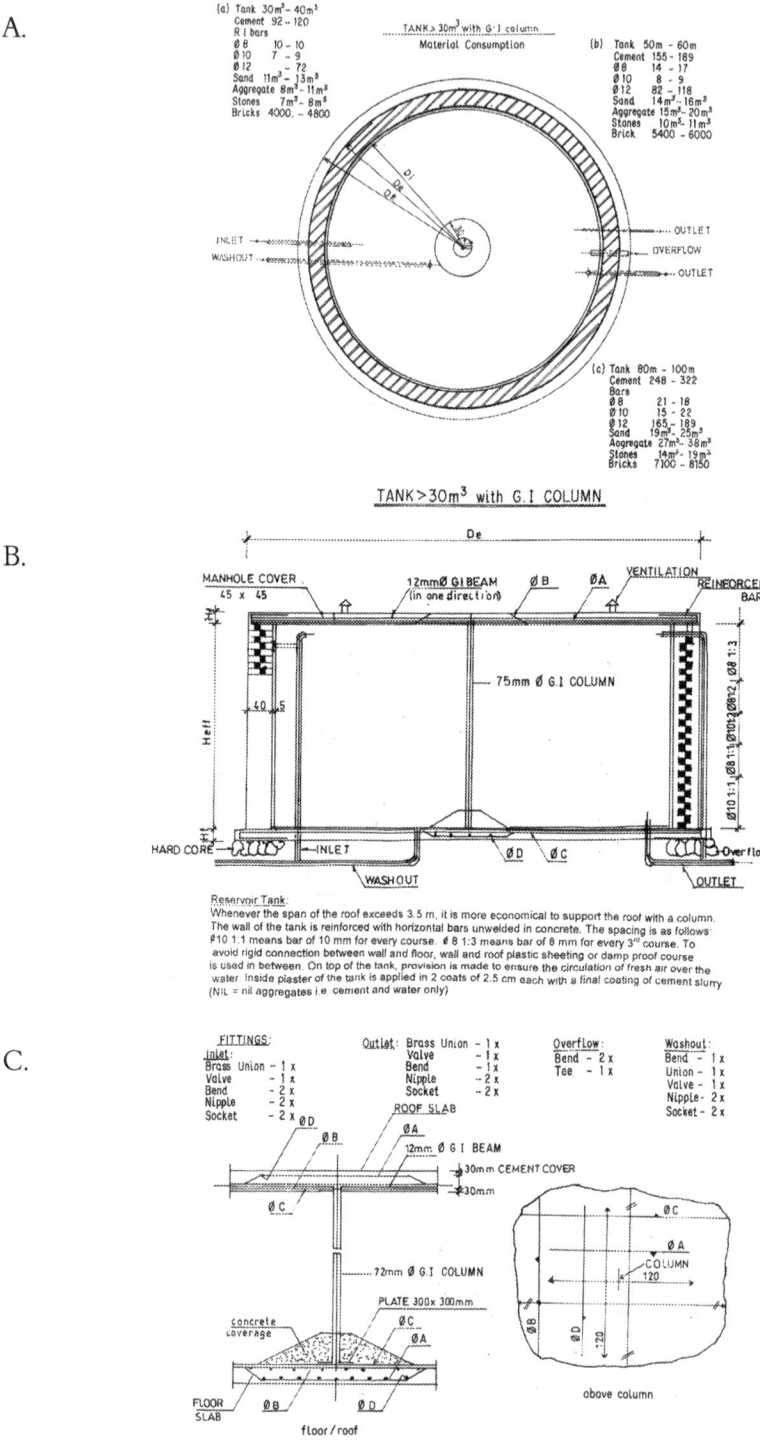

Figure 2.10A, B, C: Plan, cross-section and details of a cylindrical ground tank with diameter more than 3.5 m

The brick masonry wall is reinforced with unwedded horizontal bars embedded in concrete. The bar sizes and spacing indicated in Figure 10B are as follows:

10 1:1 means 10 mm bar in every course
8 1:3 means 8 mm bar in every third course.

To avoid rigid connection between wall and floor and between wall and roof, plastic sheets or damp proof course (DPC) is used in between.

Provision is made on the tank roof for fresh air circulation over the water – this is by means of vents shown in Figure 10B.

The inner plaster is applied in two coats of 250 mm each, finished with a final cement slurry to ensure water tightness.

2.7.2 HEMISPHERICAL UNDERGROUND FERRO-CEMENT TANK, EXTENDED ABOVE GROUND LEVEL WITH A CYLINDRICAL SECTION

The arrangement in relation to the roof is as in Figure 2.9A, while the design is as shown in Figures 2.11A & B.

The capacity of the tank shown is 80,000 litres (80 m³), and water is generally drawn from it either by handpumps and/or bucket and line, the latter method resulting in a higher water economy. Evaporation is also low. The nature of water delivery makes this arrangement suitable for schools, where pupils could forget to close the tap or even play with the running tap, thereby wasting the precious water. If the tank is full at the beginning of a 6-month dry season, the storage will yield 80,000/180 l/day = 444 l/day, enough for a community of 22 people.

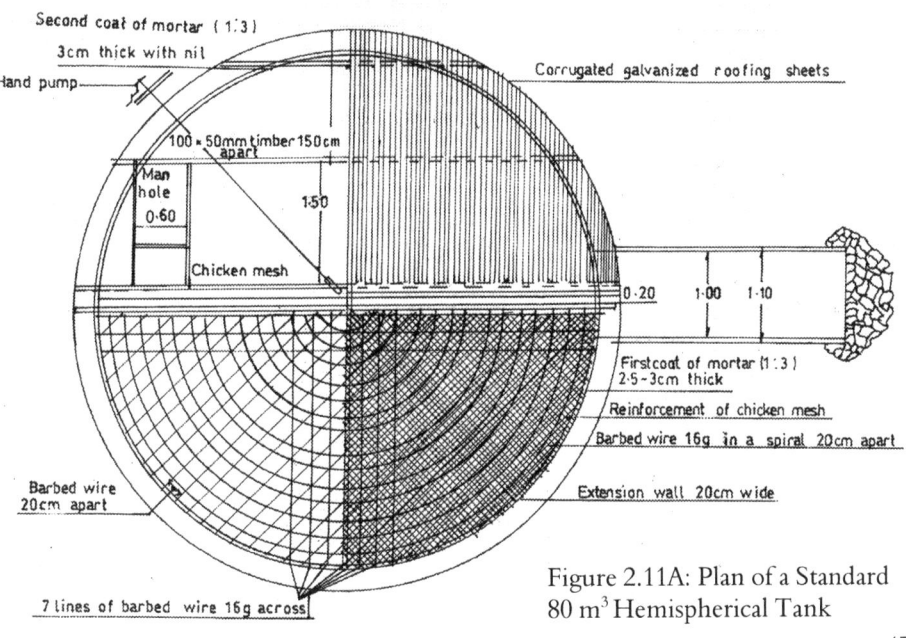

Figure 2.11A: Plan of a Standard 80 m³ Hemispherical Tank

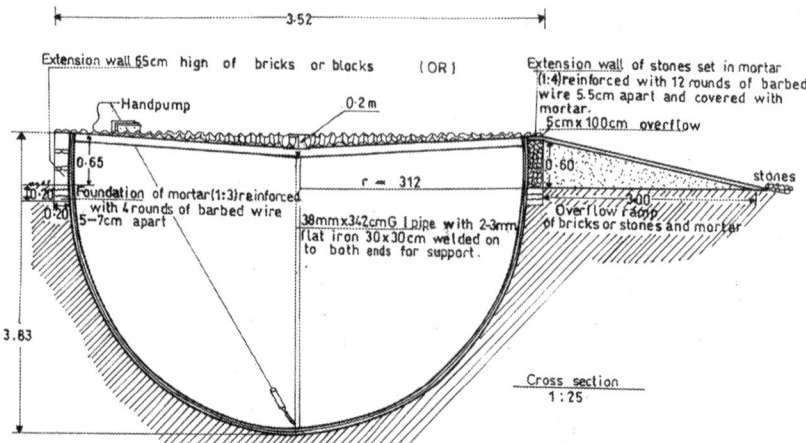

Figure 2.11B: Cross-section of a Standard 80 m³ Hemispherical Tank

The bill of quantities is given in Table 2.6.

TABLE 2.6 BILL OF QUANTITIES FOR A 80 M³ (80,000 LIT.) EXTENDED GROUND TANK

S/N	DESCRIPTION	UNIT	QTY	COST (use appropriate local rates and currency)	
				RATE	AMT
1	Cement	Bags	48		
2	Clean sharp sand	tonnes	14		
3	Clean stones	tonnes	6		
4	Water	drums	20		
5	Chicken mesh 25 mm x 1 m x 30 m	rolls	3		
6	Barbed wire gauge 16	rolls	3		
7	Nails 2/1"	kg	20		
8	Thin polythene sheets/bags for curing	metres	30		
9	G.I pipe 11/2" with flat iron ends	metres	3.25		
10	Cured hard timber 4" x 2"	metres	77		
11	Corrugated roofing sheets, gauge 30, 3 m	sheets	15		
12	Nails for Roof trusses 4"	kg	5		
13	Roofing nails	kg	8		
14	Triangular gutters, 20 cm overlap, gauge 26	metres	56		
15	Splash guards for gutters, 20 cm overlap	metres	50		

16	Hangers for triangular gutters, 3 mm	no		80		
17	Clot nails for fixing down pipes	kg		1		
18	Bitumen paste for sealing overlaps	kg		2		
19	Skilled labour	man-days		30-2 for 15 days		
20	Unskilled labour	man-days		100-4 for 25 days		
21	Transport	lump sum				
	TOTAL	-		-		

NOTE: Convert tonnes to the more familiar units as follows:
 1 tonne of cement = 20 bags
 1 tonne of sand = 8 wheelbarrows
 1 tonne of stones = 8 wheelbarrows
 1 tonne of water = 5 drums

2.8 Procedure for Constructing Ferro-Cement Tanks

2.8.1 NOTES

(a) Dimensions are given for only two surface (above ground) tanks – 10 m^3 and 23 m^3, others could be designed, depending on needs.

(b) Roof for the 10 m^3 tank is ferro-cement while that of 23 m^3 is corrugated iron sheets; these are examples, either could be used, depending on the local circumstances.

(c) The 23 m^3 storage is based on an annual rainfall of 600 mm, 50% of which is expected to yield water for storage. Thus, on a 72 m^2 roof, a run-off volume of 600 x 50/100 x 72 x 1000 l. = 21,600 l. (notice that the 23 m^3 tank is slightly bigger than the available water).

(d) Similarly, the 80 m^3 tank is based on an annual rainfall of 900 mm, 450 of which is available. Thus, on a roof of 176 m^2, a roof catchment of 450/1000 x 176 = 79,200 l (79.2 m^3) is expected. A tank capacity of 80 m^3 is therefore just sufficient for the expected volume of Catchment for the year.

(e) Apart from these clarifications, the construction procedure described below applies to all ferro-cement tanks.

2.8.2 RAISED CYLINDRICAL TANKS (FIGURE 2.9A, B AND C)

2.8.2.1 Site Preparation and Positioning the Tank

a. The tank should be sited next to the house. Measure 3.05 metres from the wall and hammer a wooden peg into the ground.
b. Tie a wire to the peg and make a small loop 205 cm from the peg, cutting off the extra wire. Put a long nail through the loop.
c. Using the nail, scratch a circle on the cleared ground around the peg, using the wire radius of 205 cm. This circle of diameter 410 is the outline of the tank foundation.
d. Dig the soil out from inside the circle to a depth of 15 cm, using a long spirit level to make sure the floor of excavation is horizontal and even.

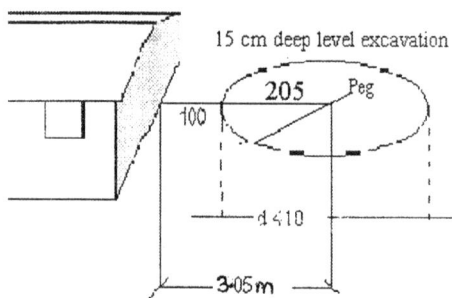

Figure 2.12A: Site Preparation for a Raised Cylindrical Tank

2.8.2.2 Preparing the B.R.C. Weld Mesh Frame

a. Cut two lengths of 400 cm long and 410 cm wide B.R.C. weld mesh. Place them next to each other with an overlap of 10cm so that the square of weld mesh so formed is 400 x 400 cm.
b. Ram a peg into the centre of the square piece of weld mesh. Tie a string to the peg and draw a circle on the weld mesh that has a radius of 200 cm.
c. Cut the weld mesh along the drawn circle line using a hammer and a chisel on hard stone. This sheet of weld mesh will be used for the foundation of the water tank.

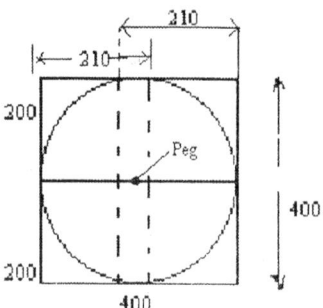

Figure 2.12B: Continuation of Site Preparation for a Raised Cylindrical Tank

d. Cut one length of B.R.C. weld mesh 1250 cm long to be used for the wall.
e. With a string, draw a circle on the ground away from the foundation that has a radius of 191 cm.
f. Place the 1250 cm long weld mesh upright as a cylinder on the drawn circle with a radius of 191 cm and tie it together with wire. The required weld mesh length = circumference of circle whose diameter = 382). The rest is the overlap required for (g) below.
g. Bend the lower ends of weld mesh's vertical ends outwards.

Circumference = πd = 22/7 x 382 = 1200
Therefore, an overlap of about 50 cm on the B.R.C. wall reinforcement. The wall B.R.C. then has a foundation B.R.C. of about 9 cm (200 – 191).

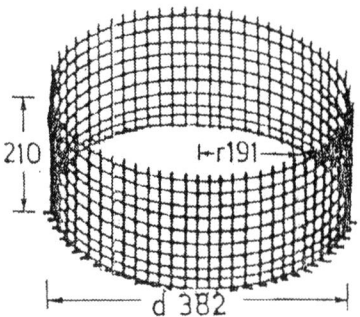

Figure 2.12C: Preparing BRC Weld Mesh Frame

2.8.2.3 Concreting the Foundation Must be Complete in One Day

a. Mix concrete made of 1 portion of cement to 3 portions of clean, coarse sand and thereafter add 4 portions of ballast (approximately 5 cm diameter stones) (1:2:4) and add water.
b. Pour a 5 cm thick concrete mix into the excavated foundation and compact it well.
c. Place the circular sheet of weld mesh on top of the moist concrete.
d. Place the cylindrical sheet of weld mesh standing vertically on top of the circular sheet of weld mesh and tie them together with binding wire in 30 places.
e. Tie the water pipe made from galvanised iron pipe onto the circular weld mesh and the wall.
f. Mix concrete (1:3:4) and pour it in a 5 cm thick layer onto the first layer of concrete in the foundation. Compact this second layer well to ensure that it has a good bond with the weld mesh underneath.
g. Keep the concrete moist and covered for proper curing.

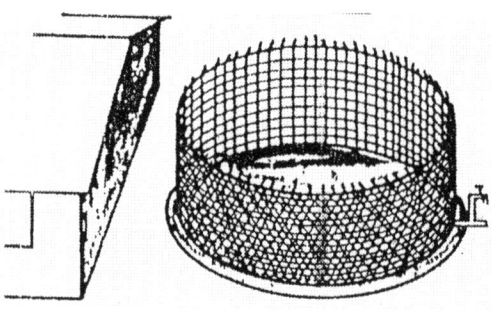

Figure 2.12D: Continuation of Figure 2.12C

2.8.2.4 Lining the Tank with the First Coat of Plaster
a. Roll one layer of 2.5 cm chicken mesh around the outside of the cylindrical weld mesh as tightly as possible. Overlapping of the chicken mesh should be at least 15 cm.
b. Roll loops of gauge 16 galvanised wire around the chicken mesh as tightly as possible. These loops of wire should be spaced 10 cm apart.

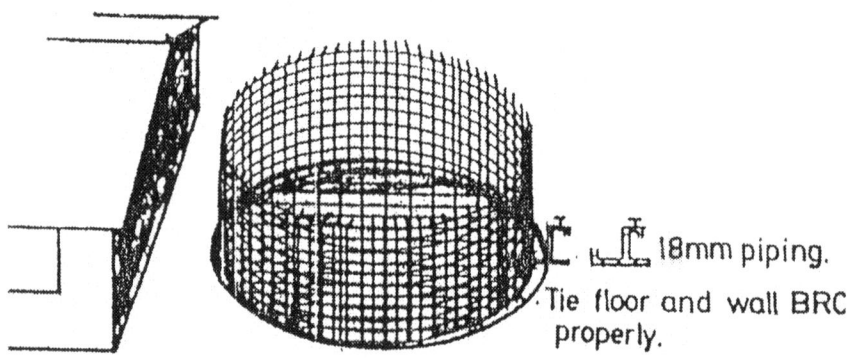

Figure 2.12E: Continuation of Figure 2.12D

c. Cut open 20 empty sugar/poultry feed sacks. Sew them together to make a long blanket 200 cm wide. This is better and cheaper than papyrus reed mats sometimes used.
d. Roll this blanket tightly around the weld mesh and the chicken mesh so that all of the outside of the tank is covered.
e. Tie 50 rounds of sisal string tightly around the blanket of sacks. Space the strings 4 to 5 cm apart.
f. Mix one portion of cement with 3 portions of clean coarse sand (1:3) and water.

g. Smear this moist mixture onto the inside of the blanket with a trowel. This coat should be 3 to 5 cm thick and stick well to the chicken mesh all over. Leave this coat of mortar to harden for 12 hours.

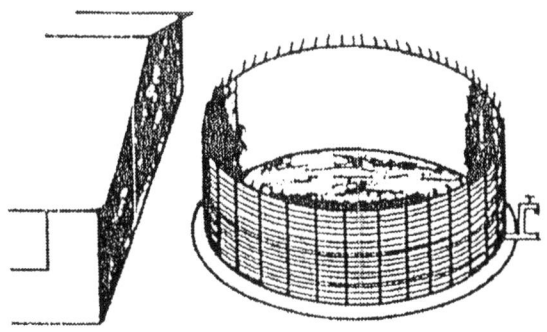

Figure 2.12F: Lining the Tank with Plaster

2.8.2.5 Lining the Tank with the Second Coat of Plaster.

a. The following day, mix mortar (1:3) and apply it to the inside of the tank in a 3 cm thick coat. Smooth it evenly with a wooden float.

b. Clean the floor of tank. Pour a 2 cm thick layer of mortar (1:3) onto the clean, moist floor. Smooth it with a wooden float.

c. Mix cement with water and apply a thin coat of this cement slurry to the moist plaster on the inside of the tank wall and floor. Press the slurry into a plaster with a square steel trowel. This will ensure a watertight coat. The coat of plaster and cement slurry must be completed in one day.

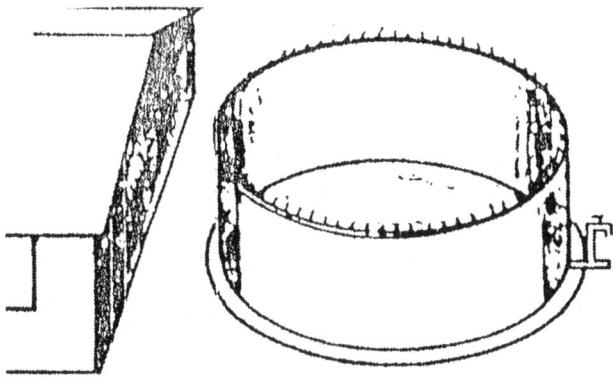

Figure 2.12G: Curing the Lining and Coating the Tank

2.8.2.6 Curing the Lining and Coating the Tank

a. On the day after lining the inside, remove the sugar-sack blanket from the outside of the tank, wet and hang it on the inside of the tank for proper curing. Lay polythene sheeting on the floor of the tank and cover it with 3 cm of water.

b. Plaster the outside of the tank with a 3 cm thick coat of mortar (1:3) and smooth it with a wooden float.
c. Wrap the outside of the tank with polythene sheets and secure it with sisal strings.
d. Keep the ferro-cement tank moist and covered for 3 weeks to ensure a good curing and a watertight tank.

2.8.2.7 Roofing

A. FIXING THE ROOFING TIMBER (OR FERRO-CEMENT AS FOR WALL)
a. Water tanks should be roofed to reduce evaporation which can take up to half the water in the tank (annual evaporation rate in arid and semi-arid (*ASAL*) regions is more than 2 metres annually, which is equivalent to the height of this particular tank).
b. Place 5 lengths of timber, 10 cm x 5 cm (4" x 2") across the tank and space them 90 cm apart. The timber should have been treated with wood preservative prior to the construction to prevent it from rotting in a moist environment. In coastal areas this can be done by soaking the timber in sea water for one month. Tie the ends of these pieces of timber to the upper part of the wall reinforcement with binding wire.
c. Place lengths of timber, 10 cm x 5 cm between the ends of the first 5 lengths of timber and nail them together with 10 cm (4") nails. Leave two gaps, 12 cm wide, for the intake of the downpipe gutters. Fix a length of timber and a frame for a manhole in the roof structure.

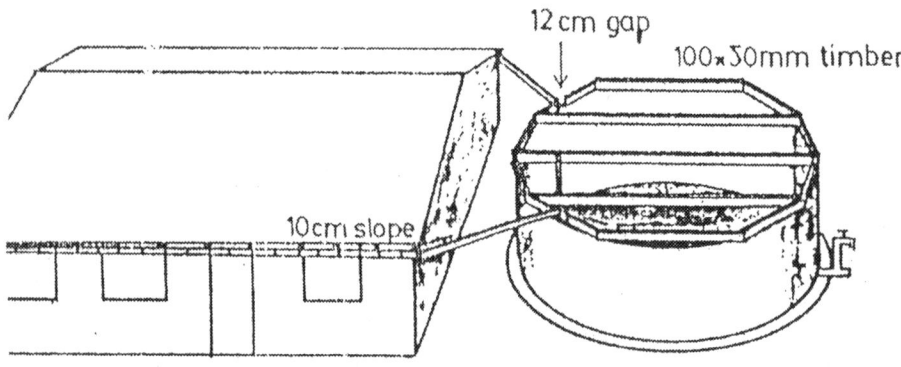

Figure 2.12H: Fixing Roofing Timber

B. CONSTRUCTING THE ROOF
a. Nail galvanised corrugated roofing sheets onto the timber structure. Trim the sheets that protrude over the edge of the tank wall. Roof the lid of the manhole and attach two strips of metal to act as handles for the manhole lid.

b. Close all openings between the iron sheets, the timber and the edge of the tank with mortar (1:4) to keep lizards and insects out.

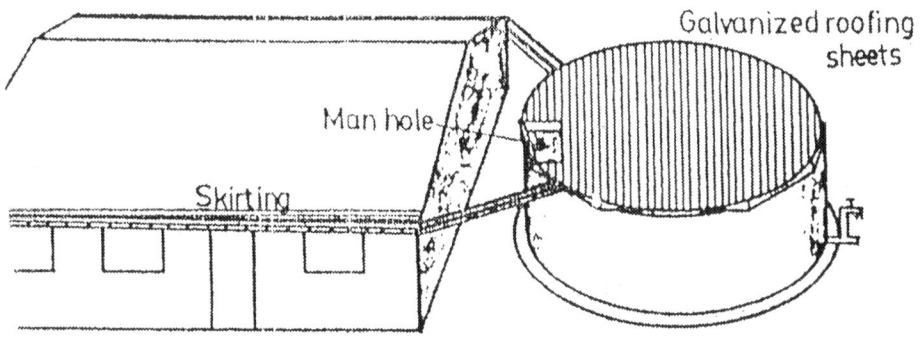

Figure 2.12I: Galvanised Roofing Sheets in Place

2.8.3 EXTENDED HEMISPHERICAL GROUND TANKS (FIGURES 2.6 & 2.11).

2.8.3.1 Site Preparation and Positioning the Tank

a. The tank should be sited next to the house. Measure 4 m from the wall and hammer a wooden peg into the ground.

b. Tie a wire to the peg, make a small loop 312 cm from the peg and cut off the extra wire. Put a long nail through the loop.

c. Using the nail, scratch a circle on the cleared ground, using the wire radius of 312 cm around the wooden peg. This circle of diameter 624 cm is the outline of the tank excavation.

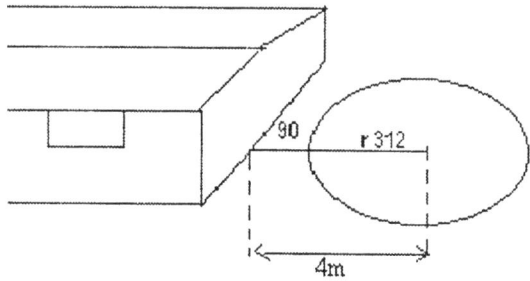

Figure 2.13A: Site Preparation

2.8.3.2 Excavating the Tank

a. Dig out the soil inside the circle leaving some soil remaining around the centre peg. Use the radius wire at all times to make the shape of the excavation half-ball. If large stones protrude out of the sides, They can be left as part of the tank wall but smaller stones should be removed.

b. After excavating all around the pillar with the centre peg, remove the pillar as the last part of the excavation.

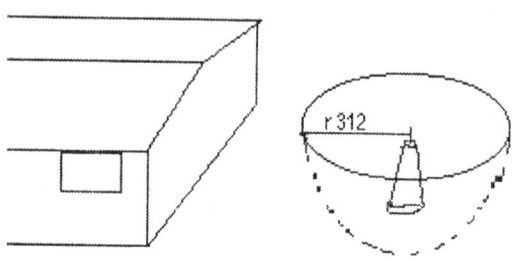

Figure 2.13B: Tank Excavation

2.8.3.3 Building instructions

A. APPLYING THE FIRST COAT OF PLASTER

a. Fill in any holes left by the stones removed during the excavation with smaller stones, mortar made from 1 portion of cement to 5 portions of sand (1:5).

b. Starting at the bottom of the tank, throw a coat of mortar, consisting of 1 portion of cement to 3 portions of coarse sand (1:3), onto the wall of soil. Continue until the plaster has reached a thickness of 2.5 to 3 cm. Leave the plaster with a rough surface as it makes a good bond for the next coat of plaster to follow in a couple of days. Keep the plaster moist and covered with polythene sheet.

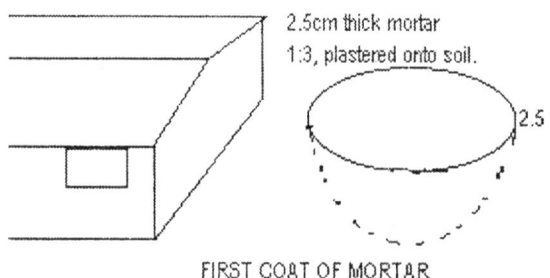

Figure 2.13C: Coating the Excavated Tank with Mortar

B. BUILDING THE TANK RIM BEAM
a. Dig a 20 cm wide and 20 cm deep ditch along the tank wall.
b. Fill the bottom of the ditch with a 5 to 7 cm layer of mortar (1:3). Place 2 rounds of barbed wire (gauge 16) on the mortar.
c. Pour a second layer of 5 to 7 cm of mortar over the barbed wire and compact it well.
d. Place another two rounds of barbed wire (gauge 16) on the mortar so that all together, there are 4 rounds of uncut wire in the ring beam. Cut the barbed wire free from its roll.
e. Fill up the ditch with mortar (1:3) and compact it well. Level the top of the beam by placing stones set in mortar.
f. Keep the ring beam moist and shade for proper cutting either with polythene sheets or sackcloth.

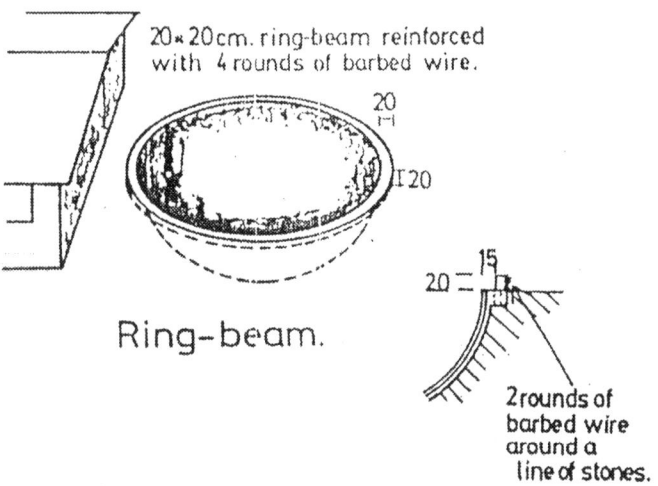

Figure 2.13D & Figure 2.13E: Constructing Ring Beam

C EXTENSION OF THE TANK WALL
a. The extension wall, 65 cm high, is built on to the ring beam.
b. A gap 100 cm wide and 5 cm deep should be left at the side furthest away from the building to allow overflow water out.
c. The wall can be built of stones set in mortar (1:4). Twelve rounds of barbed wire are drawn tightly around the wall as reinforcement and covered with mortar/plaster (1:4).
d. Alternatively, bricks or blocks with a ring of barbed wire in each course can be used.

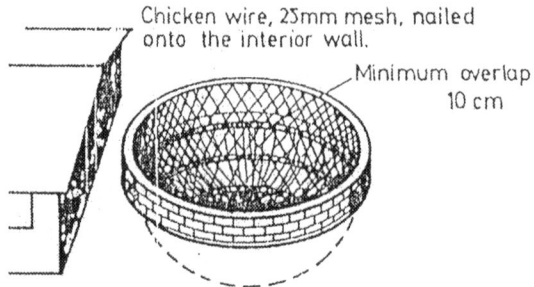

Figure 2.13F: Preparing Reinforcement for Extension Wall

D. REINFORCEMENT OF THE TANK.
a. Nail chicken mesh with 25 mm nails onto the plaster of the tank wall with 6.35 cm (2.5') nails. Use wet nails as it makes it easier to hammer them through the plaster. The overlapping of the chicken mesh must be at least 10 cm wide. If chicken mesh is not available, barbed wire can be used instead by doubling the barbed wire.

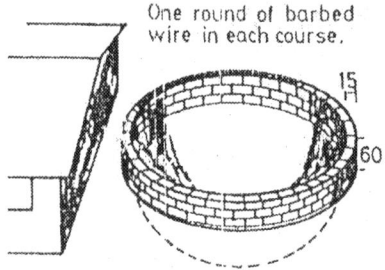

Figure 2.13G: Building the Extension of Wall

b. Nail spiral of barbed wire gauge 16, onto the interior of the tank. Start nailing the wire at the bottom centre of the tank and continue upwards to the top edge of the tank. Spacing between the lines of wire should be 10-15 apart.

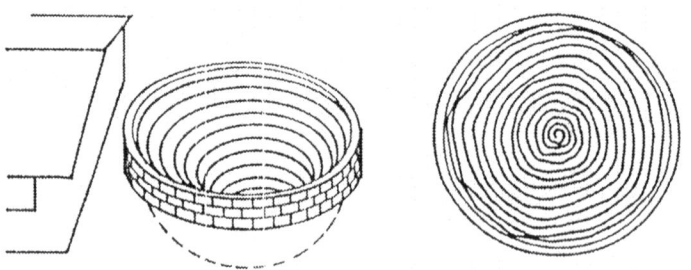

Figure 2.13H: Preparing Reinforcement for Tank Lining

c. Nail seven straight lines of barbed wire, gauge 16, across the tank passing over the bottom centre. Then nail another seven straight lines of barbed wire across the tank bottom at an angle of 90 degrees to the first seven lines. Then fix lengths of barbed wire from these lines straight up the sides of the tank as shown, nailed 20 cm apart. If chicken mesh is not available, barbed wire can be used instead, but the spacing must be reduced to 10 cm.

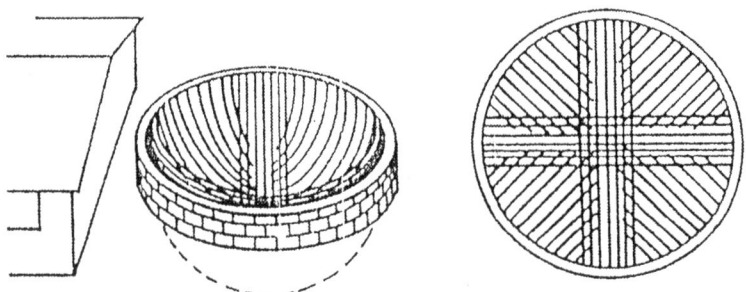

Figure 2.13I: Reinforcement of Lining (continued)

E. APPLYING THE SECOND COAT OF PLASTER

d. This second and final coat of plaster must be completed within one day to prevent cracking. Therefore, all materials, including water, should be on site before starting to plaster.
e. Clean the interior of the tank and keep the reinforcement and the wall moist.
f. Throw mortar (1:3) onto the reinforced wall, ensure that it covers the reinforcement properly. This plaster layer should be 3 cm thick so that together with the first coat of plaster, the total thickness of the wall is 5.5 cm. Smooth and finish with a wooden float until an even and uniform surface is obtained.
g. For waterproofing, apply a coat of cement slurry (cement and water mix) with a square steel trowel. Coating with cement slurry must be completed within the same day as the final coat of plaster to ensure proper bonding.
h. Cure the plastered ferro-cement work with polythene sheet and water for 2 to 4 weeks to ensure a strong structure devoid of cracks.

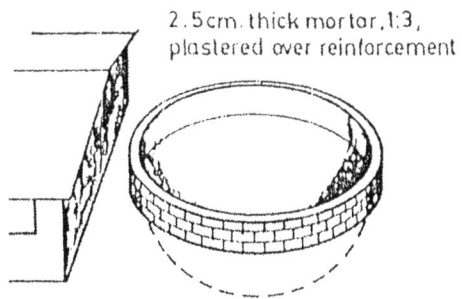

Figure 2.13J: Applying Final Coat of Mortar in Lining

F. BUILDING THE OVERFLOW
a. Build a ramp of bricks or ballast and mortar against the extension wall of the tank leading up to the 100 cm x 5 cm gap. It should be inclined on a base of 3 m x 1.10 m.
b. Plaster the surface and sides of the ramp with mortar (1:3).
c. Place two pieces of timber 10 cm x 5 cm on the ramp, 5 cm in from either edge as a form-guide. Fill this 5 cm between the edge and the timber with mortar to make the sides of the overflow chute.
d. Place an apron of stones at the lower end of the ramp and dig a ditch down to a hollow planted with fruit trees.

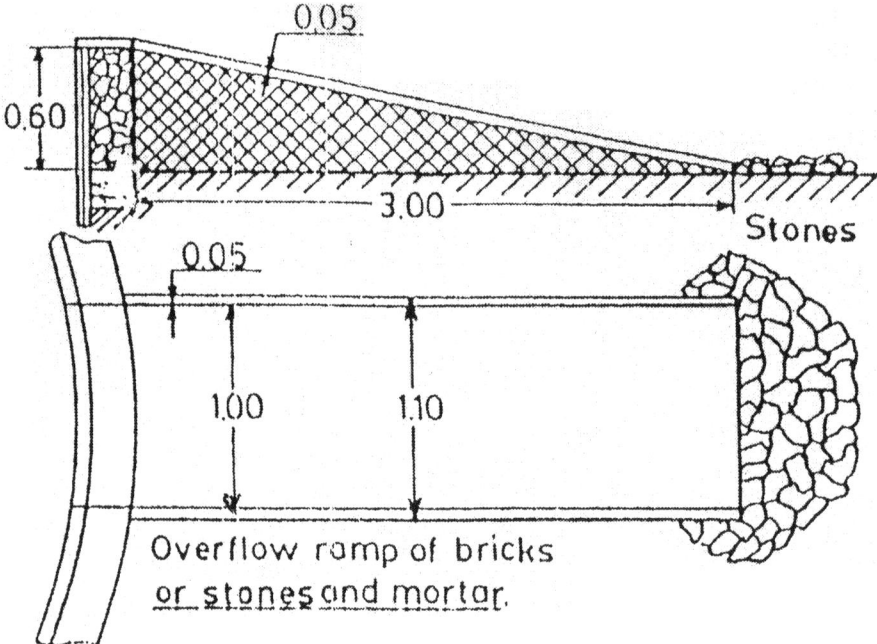

Figure 2.13K: Building the Tank Overflow

2.8.3.4 Roofing

A FIXING THE ROOFING TIMBER.
a. Place a 3.75 cm (1.5") galvanised iron pipe with flat pieces of iron welded to both ends. Place the pipe upright in the tank with the lower end at the centre of the tank bottom. The upper end of the pipe must be about 20 cm lower than the top of the extension wall in order to make the tank roof slope towards the centre of the tank. See Figure 2.10.B.
b. Place 2 lengths of 10 cm x 5 cm (4" x 2") timber from wall to wall, resting on the flat iron welded to the centre pipe. All timber should have been treated with wood preservative as for cylindrical tank roof. These 2 timber members should be placed about 20 cm apart. Nail 2 layers of chicken mesh between the 2 timbers members to act as a sieve.

c. Place more 10 cm x 5 cm pieces of timber on either side of the first two and space them 150 cm from each other.
d. Place another 2 lengths of 10 cm timber on each side and 150 cm from the last two, spacing them 150 cm from the first pieces.
e. Use stones, blocks, bricks and mortar to build up the extension wall so that it reaches the top of the timber.
f. Make a square manhole measuring 60 x 60 cm between one set of the timbers nearest to the extension wall.

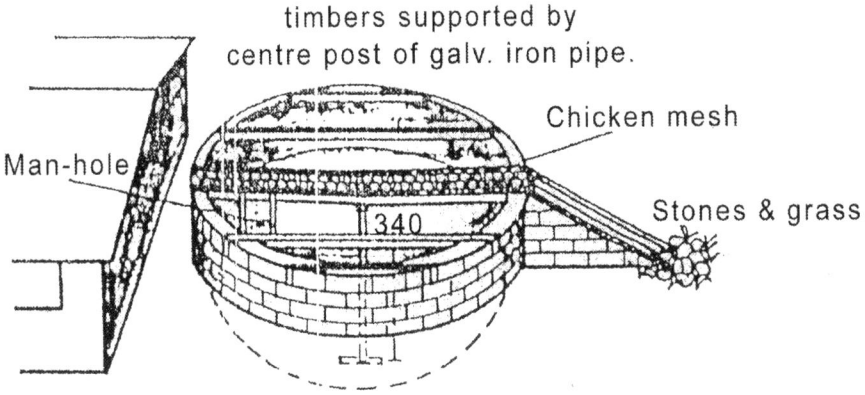

Figure 2.13L: Preparing Framework for Roof and Spill-over Ramp

B. PUTTING ON THE ROOFING SHEETS.
a. Nail corrugated, galvanised iron sheets onto the pieces of timber. Leave a space of 10 cm between the end of the sheets above the centre to allow run-off water to enter the tank through that space which is covered with chicken mesh.
b. Cut a hole in the roofing sheet for the manhole. Make a timber frame of 10 cm x 5 cm (4" x 2") and cover with the cut piece of roofing sheet to serve as cover for the manhole.
c. Trim the roofing sheets along the outer edge of the tank. Fill all the spaces between the sheets and the tank wall with mortar (1:4) to keep out lizards, insects etc.
d. Install a hand-pump next to the manhole.

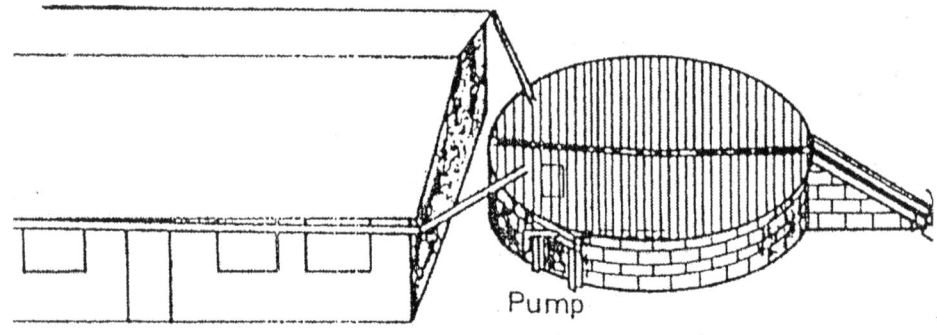

Figure 2.13M: Placing Roofing Sheets, Handpump and Gutters

2.8.3.5 General Quality Control.
There are some general guidelines that must be followed to ensure quality control and the production of leak-proof tanks.

A. MORTARING, PLASTERING AND CURING.

a. All coats of plaster must be completed within one day.
b. The plaster mix and reinforcement at all stages must comply with specification. Note that any reduction in cement in the mix, either through a false attempt at economy or through cheating will make the wall crumble and flake after wetting.
c. Similarly if the reinforcement is not properly made, large cracks are likely to appear in the tank. If the curing is not carried out effectively, many small cracks will appear.
d. The polythene sheet cover should remain over the lining for long enough, and the mortar kept moist to ensure that the lining does not dry too quickly to encourage the occurrence of cracks.
e. However, if cracks appear in the tank, depending on the seriousness of the cracking, several actions can be taken.
   i. If the lining crumbles off the tank wall, the mortar needs to be removed and the lining redone from scratch.
   ii. If the surface is covered with small cracks, this may be due to too rapid drying; a new coat of plaster can be applied with a second waterproofing. Curing should be done properly this time.
   iii. If there is one or two large cracks through which water leaks out of the tank, these need to be blocked.
a. The cracks need to be chiselled out and the leakage blocked. The best

method is to chisel away 20 cm on either side of the crack until the reinforcement is exposed. Thereafter, nail 30 pieces of 5 x 5 cm (2' x 2') weld mesh 120 cm long across the cracks. The weld mesh pieces must overlap at the ends by at least 15 cm. Then clean the exposed area to be re-plastered with water and throw dry cement dust onto the moist surface. Allow this pure cement coat to dry for about 4 hours before applying plaster, i.e. mortar (1:3). Finally apply a smooth complete coat of plaster over the crack and onto the wall of the tank around the crack.

b. Small cracks can be painted with bitumen paste instead of this elaborate patching.

B. PRECAUTIONS ON ROOF GUTTERS

a. The roofing on the tank should be constructed so that it drains towards the centre of the tank so that rainwater may flow in properly.

b. The gutters should be hung properly on the roof edge so that they slope gradually down towards the tank to ensure that the water running off the roof will flow into the tank

2.8.3.6 Cleaning and Maintenance

a. This being rainwater free from pollution, it is already of reasonably good quality, compared with water from natural depressions.

b. To maintain this water quality, the gutters and the roof should be cleaned periodically, especially just before the outset of the rainy season to prevent leaves, dust, dead animals and insects from being washed into the tank. There should be a chicken wire and screen-mesh filter over all the inlets and the tank overflow to keep out pollutants and stop insects from entering. All or part of the first rain of the season should be diverted by moving the down-pipe so that the roof washout may not flow into the tank.

c. If the tank is not empty at the end of the dry season, it should be emptied and any silt collected at the bottom removed. The interior of the tank should be cleaned with a brush.

d. Any gaps noticed between the roof and the extension wall, should be filled with stones and mortar to keep out wind-blown materials, lizards, birds and insects. This will also minimise evaporation losses.

e. For the extended ground tank, a clean and efficient method of drawing water from the tank can be adopted, the best being a handpump fixed next to the extension wall with its suction pipe fixed just above the bed of the tank (see Figure 2.11B). On the ground around the handpump, a square stone platform of about 1.2 m square should be laid to prevent the ground from getting swampy from spilled water. The handpump should be locked with a padlock to prevent unauthorised use.

Chapter III
GROUNDWATER

3.1 Occurrence of Groundwater (subsurface water)

Groundwater is the portion of rain (or atmospheric precipitation), which, after infiltration, percolates (penetrates) into the underlying earth strata. The underground strata that bears (stores) water is called an aquifer. The amount of water that can be stored in an aquifer depends on the porosity of the aquifer. Porosity of a formation is the pore space between its particles, it is the volume of the voids. A good aquifer (one with high porosity) is made up of unconsolidated geological materials like sands, clay, topsoil, gravels and glacial drift, while aquifers made up of solid formations like limestone, granite and gneiss do not hold much water and are said to be poor aquifers. Ground formations are known to have a tendency to hold the water and release only part of it. The permeability of a formation is a measure of its ability or tendency to give up or release part of the water it holds. Water passes through large openings more easily than it does through small openings. Thus, clay and topsoil have high porosity (large volumes of voids) but low permeability (small openings between particles) and will not allow water to pass freely. On the other hand, gravels, sands and sandstone have high permeability and will allow water to pass with relative ease. The specific yield of an aquifer is the ratio of the water which will flow freely from the material to the total volume of the formation; it is always less than porosity and is a more practical way of conceptualising how good an aquifer is in terms of water provision than permeability. For purposes of groundwater development, sands, gravels and sandstone, which also have high porosity are the most amenable and therefore of greatest interest.

Permeability is a function of porosity and structure (i.e. grain size, distribution, orientation, arrangement and shape) of the material, as well as its geological history. Details are beyond the scope of this book – see references for further reading. Permeability of a material is defined by its permeability coefficient k, (also called hydraulic conductivity) which depends on the factors listed above, it is measured in metres per day.

Table 3.1 gives the porosity, specific yield and permeability coefficient of commonly encountered naturally occurring soils and rocks.

TABLE 3.1   POROSITY AND PERMEABILITY COEFFICIENT OF SOME SOIL FORMATIONS

SN	FORMATION	POROSITY (%)	SPECIFIC YIELD (%)	PERMEABILITY COEFFICIENT k (M/DAY)
1	Sands and gravels of fairly uniform size and moderately compacted	35–40	} 16	} gravels 10^3–10^5 } fine gravel 10^2–10^3
2	Well-graded and compacted sands and gravels	25–30	} Sand 25 Gravel 22	} coarse sand 10^1–10^2 } fine sand 10^0–10^1 } very fine sand 10^{-1}–10^0
3	Sandstone	4–30	8	
4	Chalk	14–45		
5	Granite, schist and gneiss	0.02–2	0.5	
6	Slate and shale	0.5–8	}	
7	Limestone	0.5–17	} 2	} sandy clay 10^{-4}–10^{-2}
8	Clay	44–47	3	} pure clays 10^{-6}–10^{-4}
9	Topsoil	37–65		
10	Silts	up to 80		10^{-2}–10^{-1}

SOURCE: ADAPTED FROM WILSON & LINSLEY JR ET AL.

Groundwater is subject to gravity and so it moves down the strata at rates which depend on the permeability of the strata; this downward movement continues until it reaches an impermeable stratum. When this happens, the water continues to fill the pores of the aquifer until they are full; at this point, the aquifer is said to be saturated. The surface of groundwater saturation is called the groundwater table or phreatic surface. This surface rises and falls, depending on the status of groundwater, itself determined by the weather – it rises in the rainy season and falls in dry weather. The aquifer which holds this water is called a water-table aquifer.

Except for unusual geological features or the occurrence of groundwater dams in any given drainage basin, groundwater moves slowly towards the nearest free water surface (or principal open water bodies) like rivers, lakes or even sea. Where an underlying impervious layer outcrops or is very close to the surface along the path of the groundwater, the water comes out to the surface as seepage or spring.

In some cases, the aquifer is also overlain by an impervious stratum, which puts the groundwater under pressure. This kind of aquifer is called a Confined Aquifer, and is generally fed from some considerable distance away. If a well or a hole is dug through the overlying impermeable stratum, the groundwater will rise under pressure to a point called Piezometric Surface. A well dug into this kind of aquifer is called an Artesian Well; the aquifer is also sometimes called an Artesian Aquifer.

The occurrence of groundwater as described here is represented in Figure 3.1

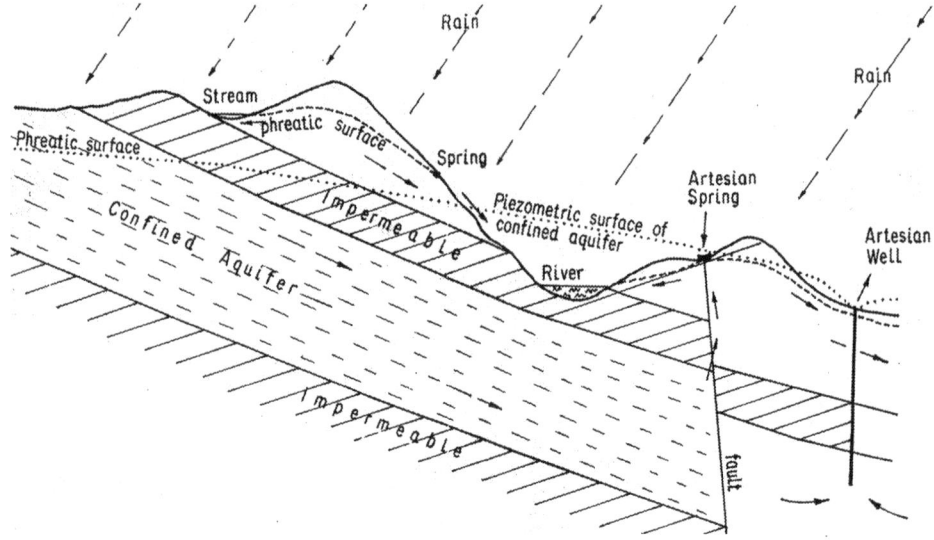

Figure 3.1: Modes of Occurrence of Groundwater

3.2 Abstraction of Groundwater

3.2.1 EXPLOITING GROUNDWATER RESOURCES

Groundwater is generally exploited by methods which bring out the water stored below the earth's surface; it is also primarily for human and livestock consumption, except for groundwater dams and in very dry areas, where crop has to be produced at all costs, in which case, groundwater is used for irrigation as well.

The most common methods of exploiting groundwater resources are:
  i. Spring Protection
  ii. Wells drilled into the water–bearing stratum (aquifer)
    - Shallow Wells
    - Tube Wells
    - Boreholes
  iii. Groundwater dams
  iv. Utilisation of high water table in lowland areas (dry farming)

3.2.2 SPRING PROTECTION

Springs are outcrops of groundwater and usually appear as small waterholes or wet spots at the foot of hills, along slopes and along river banks or other ditches. By another definition, a spring is a place where a natural outflow of groundwater occurs. Having passed through several layers of soil after infiltration, spring water is generally of good quality for domestic purposes without treatment, provided the spring is properly protected against pollution or contamination from outside. Springs therefore offer the best source of water supply for rural communities.

Generally, springs are of two types – gravity and artesian.

Gravity springs, which could be depression or overflow springs, occur in unconfined aquifers and usually have small yields, reduced further in dry weather or by the lowering of the water table in the area due to groundwater withdrawals – see Figure 3.2.

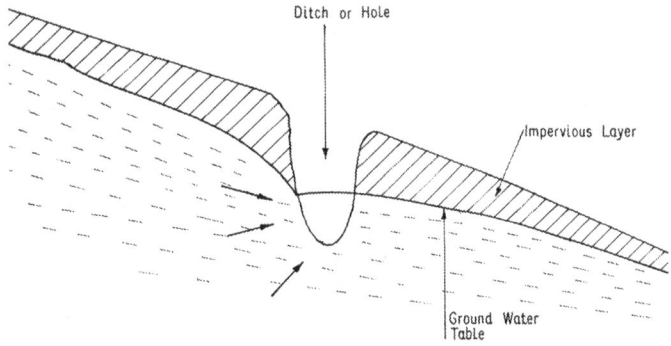

Figure 3.2A: Gravity Depression Springs

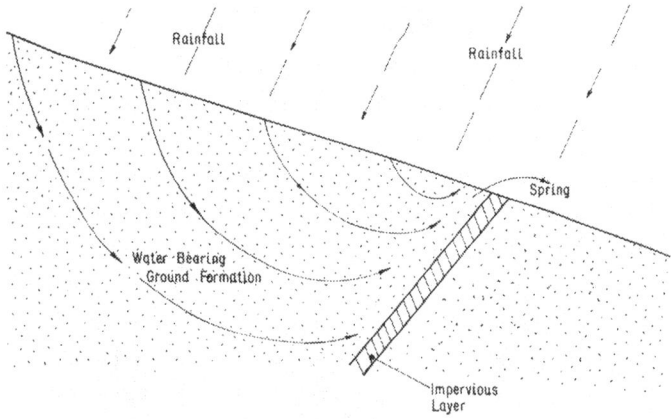

Figure 3.2B: Gravity Overflow Spring

The overflow type is as a result of the outcrop of an impervious layer – solid rock or clay – which prevents the downward flow of the groundwater and forces it to the ground surface (see Figure 3.2B). This yields much more water than a depression spring because all water from the tributary recharge area is discharged at the spring and will be more regular than from normal recharged from rainfall.

Artesian depression springs are similar in appearance to gravity depression springs. But the groundwater is prevented by an overlying impervious layer from rising to its free water level and is therefore forced out under pressure, which makes the discharge higher and subject to less fluctuation. As Figure 3.3A shows, a drop of the artesian water table, as occurs during dry periods, hardly has any influence on the yield of the spring. Artesian springs have the further advantage that the overlying impervious layer protects the water in the aquifer against contamination/pollution, with the result that the water will be bacteriologically safe.

Artesian fissure springs and Artesian overflow springs (Figures 3.3B and C) are variants of artesian depression springs, and where they occur they are good sources of community water supply.

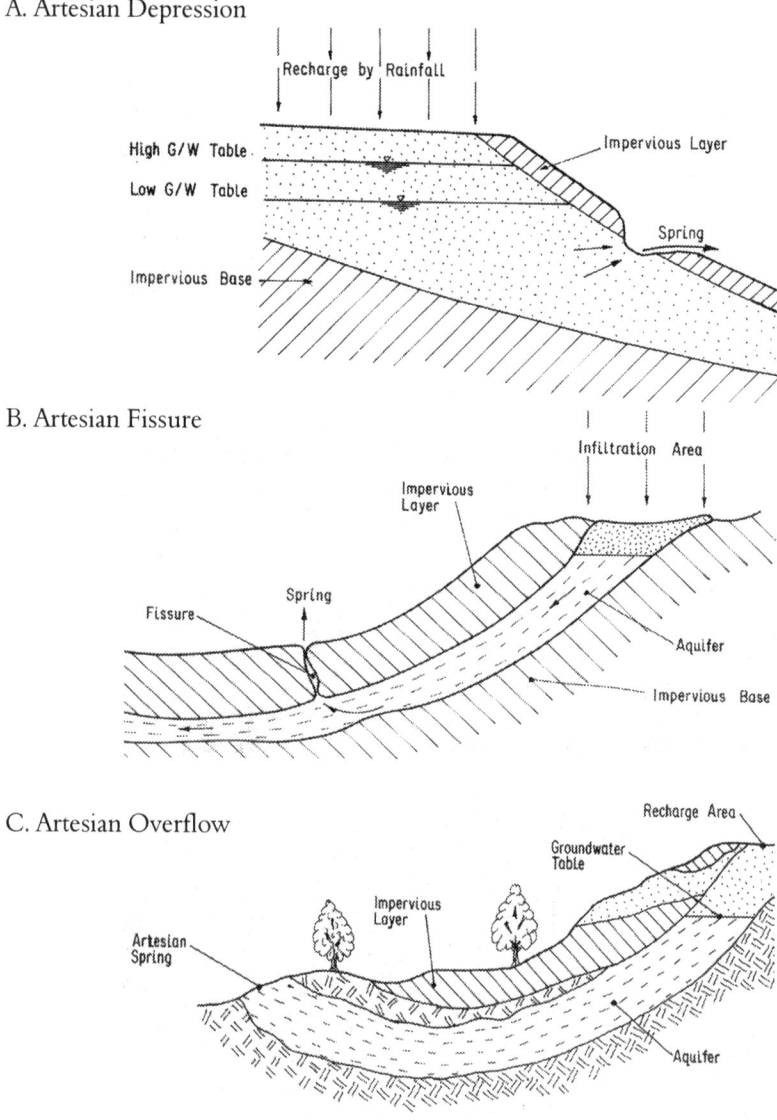

Figure 3.3: Artesian Springs

From the foregoing, it is seen that springs in general and gravity springs in particular, are prone to contamination in the area close to the point of emergence. Protection against contamination, in particular, is the main thrust of any effort to develop springs as sources of water supply. By another definition, spring protection means, essentially, concentrating the seepage and small discharges into one outlet from where it can be channelled (or even piped) to the point of use – irrigation, human and livestock watering and even recreation. Before a spring is finally selected for protection, thorough investigation should be carried out to include information on:

- origin of the groundwater
- the nature of the aquifer
- quality of the water
- its yield in various seasons of the year(rainfall)
- topography and vegetation of the surrounding area
- presence of possible sources of contamination.

The following should be noted in the course of the investigation:

- if the water differs in temperature between day and night, its quality is suspect.
- in granular aquifers, the outflow will vary little with distance along the contour (seepage springs); to tap this kind of spring, infiltration galleries of considerable length will be required, albeit their location is not critical. With fractured rock aquifers on the other hand, the flow will be concentrated where water-carrying fissures reach the ground surface. In this case, small-scale Catchment works may be sufficient; however, they need to be sited carefully.
- an assessment of the yield and seasonal variation is necessary; the result obtained will be little influenced by the spring protection works.
- the tapping of spring water has the advantage that the natural groundwater table will normally be lowered very little, if at all.

To protect a spring, it is important to note the following:

- the protection structures should be so located that any surface water infiltrating will pass through at least 3 m of soil before reaching the groundwater
- animals and humans should be excluded from at least 90–100 m distance from the protection works, e.g. by fencing.
- a diversion ditch should be dug above and around the protection works to intercept surface run-off and divert it away from the groundwater collection zone.

- Springs which can be piped to the user by gravity, often provide an economical and safe solution to the water supply problems of rural communities.

3.2.3 METHODS OF COLLECTING WATER FROM SPRINGS.

3.2.3.1 Gravity Springs

Gravity springs are particularly prone to contamination in the area close to the point of emergence. The water extraction method should take this into account. A typical gravity spring protection method is illustrated in Figure 3.4.

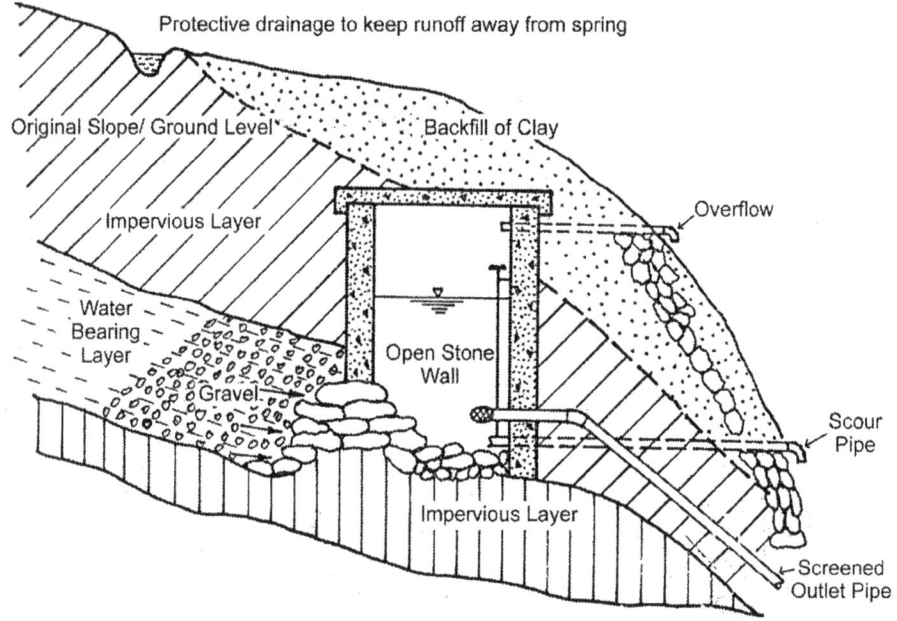

Figure 3.4: Gravity Spring Protection Method

The quantity of water from a spring can be substantially increased by digging out the area around it down to the impervious layer to remove decomposed rock, rock fragments, silt, organic matter, mineral matter some times deposited by the groundwater and other debris. Care must however be taken not to disturb the underground formations to the extent of making the spring deflect to other directions or into other formations (i.e. do not disturb the eye(s) of the spring).

3.2.3.2. Artesian Springs

Figures 3.5A, B, C and D show the protection works for different commonly encountered artesian springs, depending on the field conditions and the size of the community to serve.

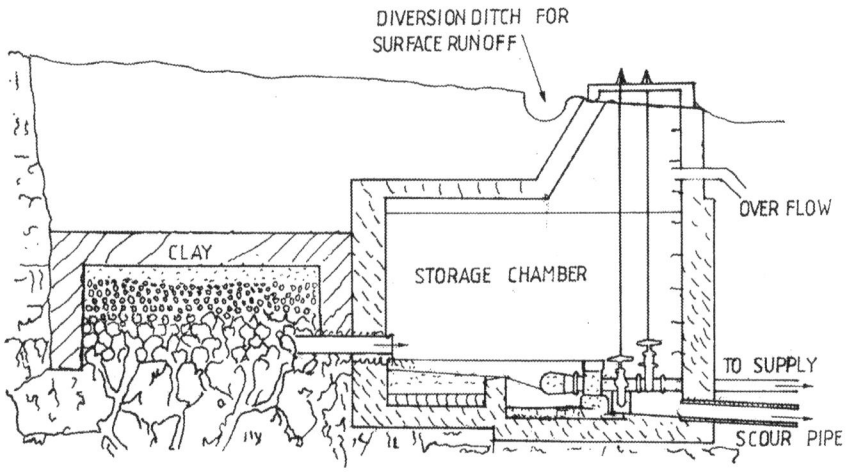

A: Depression

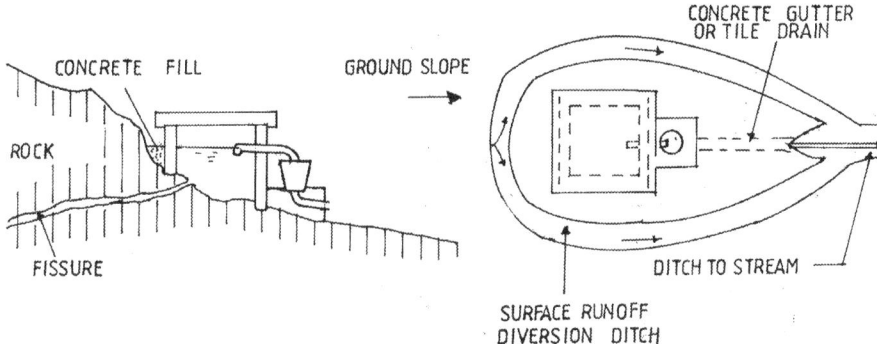

B: Fissure of Small Capacity

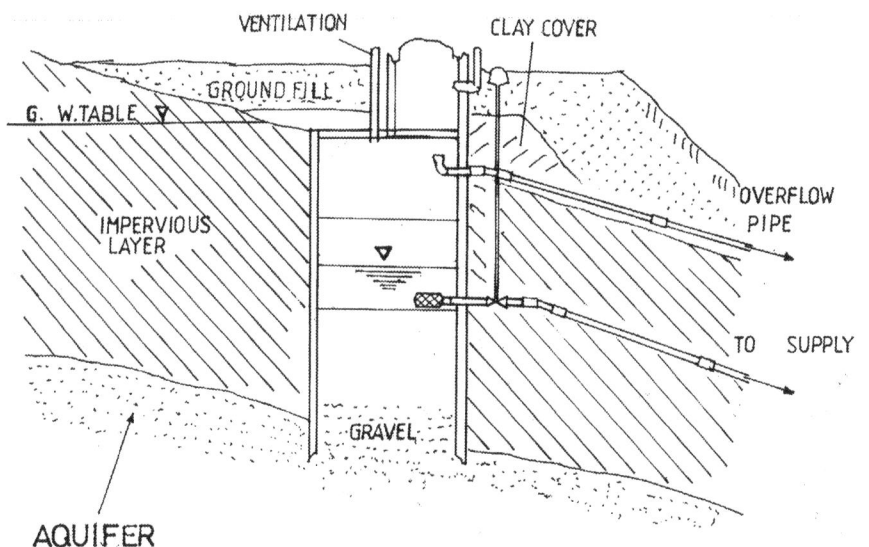

C: Fissure of Large Capacity

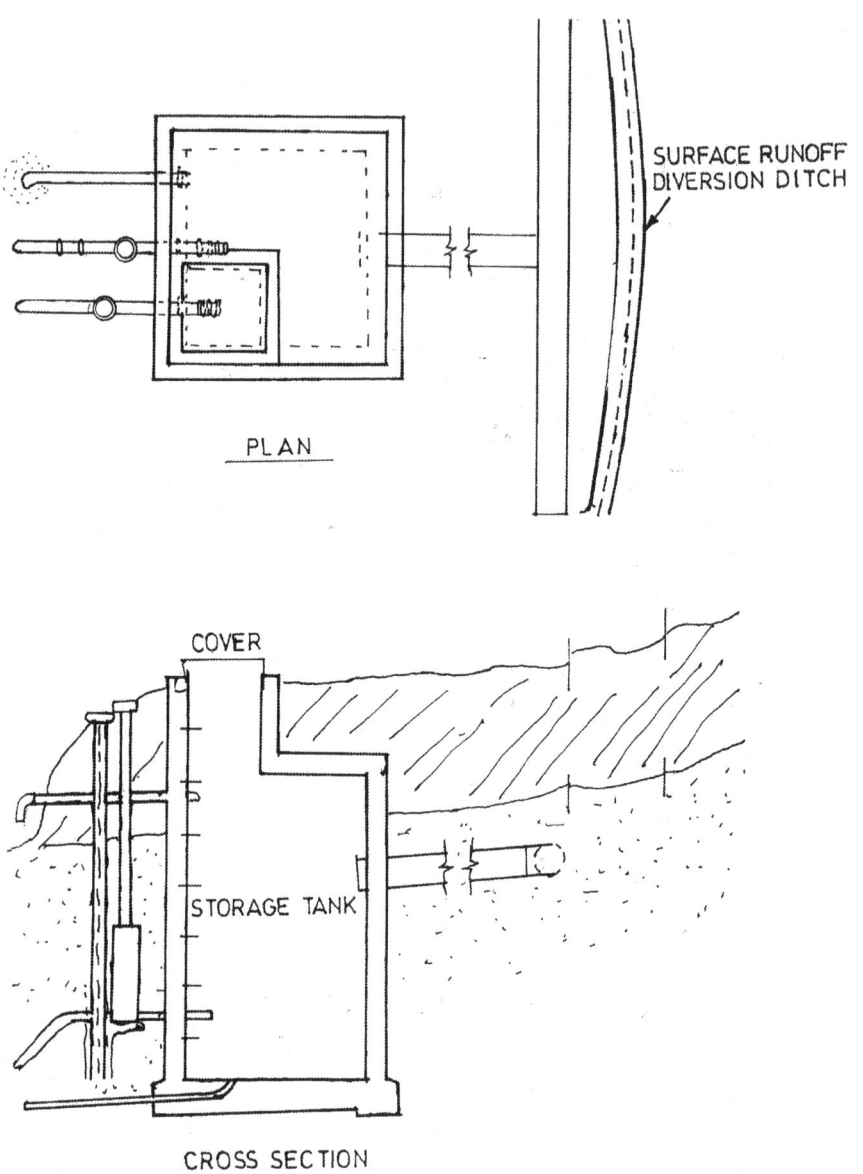

D: Contact of Large Lateral Width

Figure 3.5 Artesian Spring Protection Works

3.2.4 WELLS

The two common types of wells are Dug Wells and Drilled Wells, commonly called boreholes. In both cases, the aquifer is of considerable thickness and is usually deeper than 6 m below the ground surface. Other types like Driven Wells and Jetted Wells are uncommon and will not be considered in this handbook.

3.2.4.1 Dug Wells

Circular wells are preferred to square or rectangular wells for reasons of strength and economy. The diameter is a compromise between economy and the convenience of digging. Large diameters, which are very convenient for digging, require thicker linings, which make them quite expensive. It has also been found that the efficiency of two people working together in sinking a well is greater than twice that of a single person; two people require a minimum working space of about 1.3 m diameter. Thus a diameter of 1.3 m has more or less been standardised for most dug wells. The shaft of the well also serves as a temporary reservoir for the well water before it is withdrawn for use.

The type most commonly used is the hand-dug well, which employs several methods of sinking (excavation or digging), the basic task being to dig far into the aquifer, preferably through its entire thickness, and to protect the 'constructors' and the end users against collapse or caving in or falling/slipping in respectively. The finishing of the well also guards against pollution of the water from possible surface influent.

By far, this method seems to be about the cheapest way of providing safe water in relatively small quantities for small rural communities like villages and individual homes, especially in the absence of springs.

Lining is provided to keep the walls from caving in and when it projects above the ground level and the backfill covered with a sloping concrete slab, it also serves as a sanitary protection against pollution. The projected lining is also provided with a cover, which has openings for a pump and manhole for inspection and repairs etc. Figure 3.6 shows schematically the cross-section of a lined dug-well, protected against pollutants. Lining could be with brick or stone masonry, prefabricated concrete or clay tiles or *in situ* concrete.

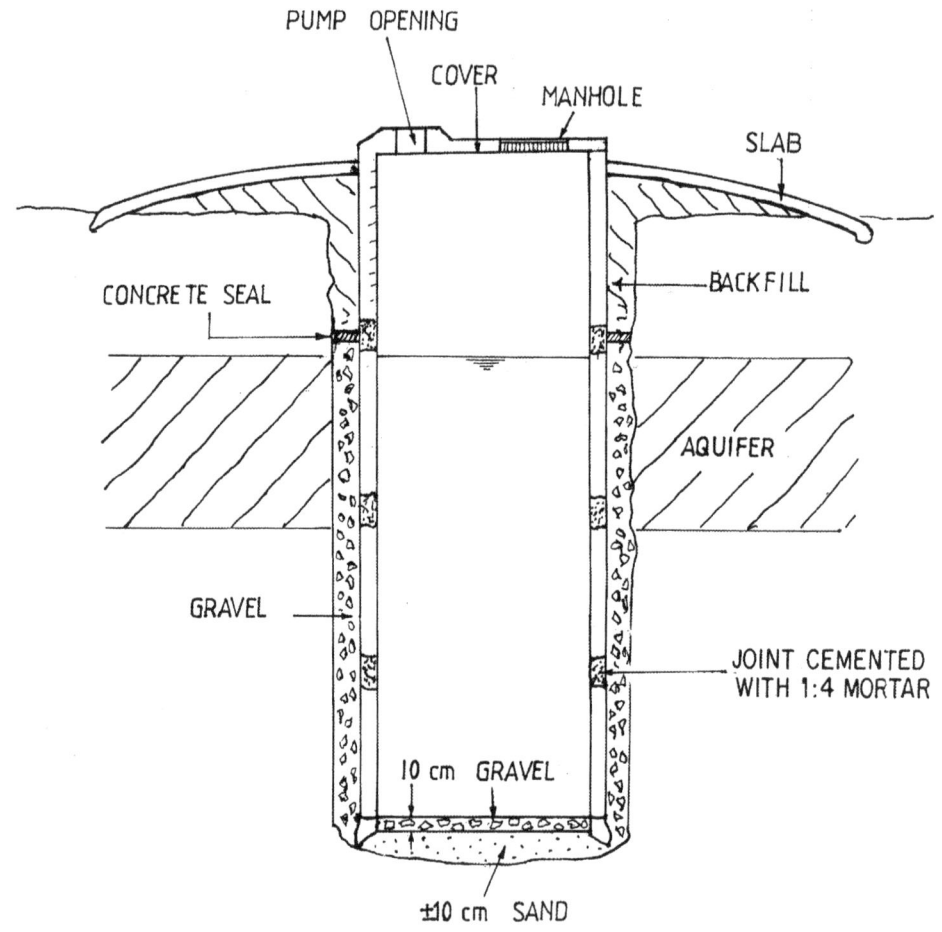

Figure 3.6: Cross-section of a Lined Well Protected Sanitarily

Water delivery from these wells is by line and bucket in very poor communities, and pumps – hand or electrical – in more affluent areas and homes. Figure 3.7 illustrates a well protected with a brick masonry lining (sanitary protection) and fitted with a pump.

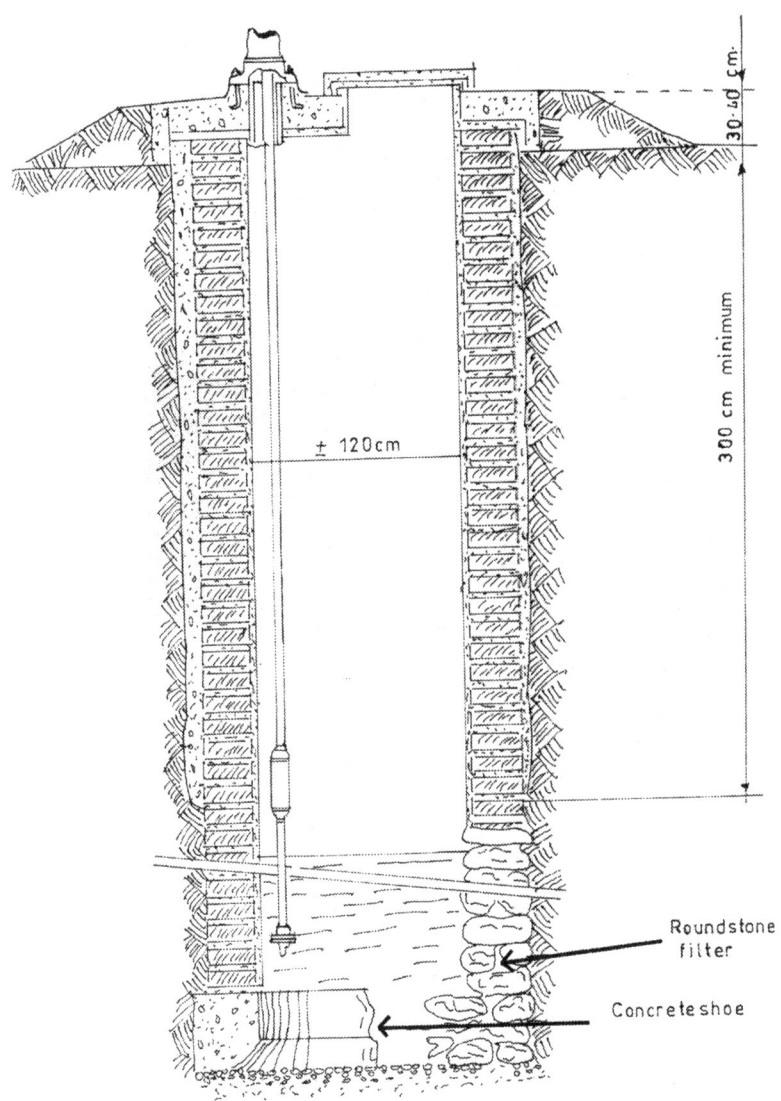

Figure 3.7: Brick Masonry Lining – notice the round stones which serve as filter and the concrete shoe on which the lining is built.

3.2.4.2 Construction of Dug Wells

Where the overburden (soil above the aquifer) is of stable structure (e.g. coarse sand and gravel), the well is sunk (dug) to its desirable depth before lining. In this situation, about half the bottom wall of lining is built on a foundation of washed round stones opposite the producing aquifer. This stone wall serves as a filter to keep out sand and other aquifer products which could block the flow from the aquifer and clog the pump. The other half is built on a concrete shoe above which the pump is eventually installed. The lining is then built up to about 30 cm or more above the ground level with bricks or concrete. Washed graded sand filter can also be laid on the floor of the well as additional protection for the pump and overdrawing.

An alternative lining method is by the use of prefabricated (prefab) concrete or clay tiles (rings), which can be quickly lowered one after the other, the lowest one resting on the concrete shoe and stone filter described above. Figure3.8 illustrates this.

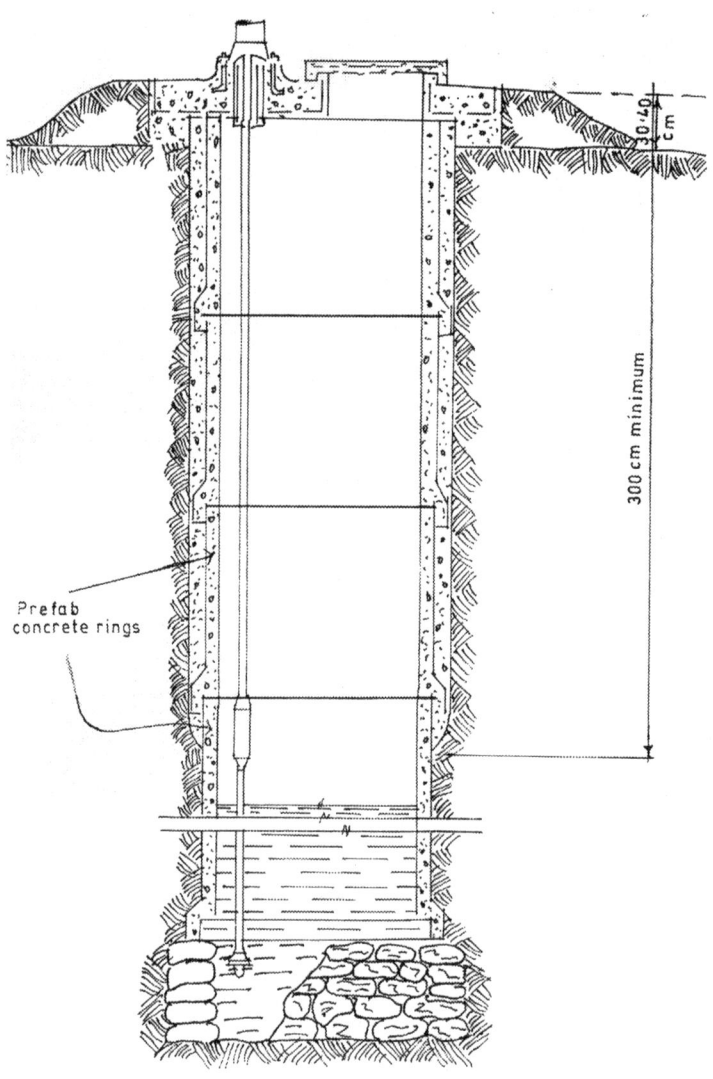

Figure 3.8: Completed Well with Pre-fabricated Concrete Lining (Rings)

The technology of this construction method is low and is usually easily transferred to unskilled workers. Also, lining with prefab concrete rings is enhanced by its ready availability in small towns near the rural communities. The excavation is a rather slow process and the equipment used simple, making it suitable in remote and relatively inaccessible locations. Figure 3.8 also illustrates a completed well with prefab concrete rings in place.

Where the well is deep, and especially where the overburden is of unstable materials, the masonry lining casing is started after the excavation is some 3–6 m deep, depending on soil conditions. The lining is built up to ground level and henceforth, it descends under its own weight as the excavation progresses. This provides protection for the workers and the rather fragile well wall (see Figure 3.9).

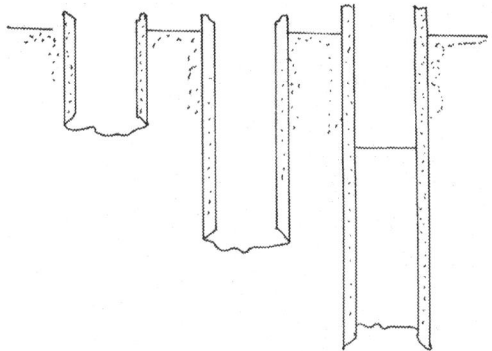

Figure 3.9A: Concrete Lining Descending under its own Weight as Excavation Progresses

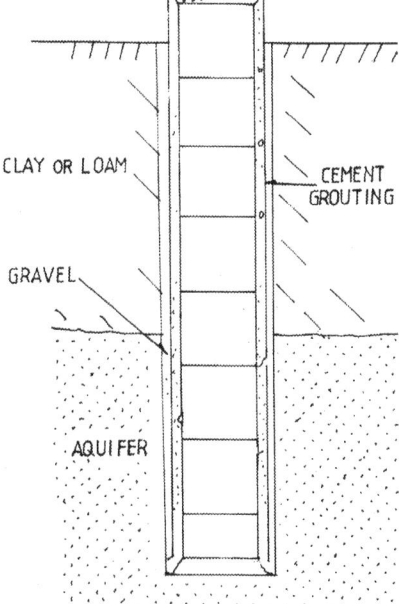

Figure 3.9B: Dug Well Construction with Prefab Rings

Where this excavation and lining are intended to be done simultaneously, the lower end of the starter ring is provided with a shoe which has a sharp inner edge for cutting and slightly wider outer diameter to facilitate sinking (excavation) and to reduce friction. This ring leaves a space around the curb, which in loose formations is self-sealing, but must be filled with cement grout or clay puddle in other formations to keep out polluted surface water. Over the thickness of the aquifer, filter rings called 'no-fines' concrete are used to allow groundwater to flow into the well. The rings are called no-fines because in the course of moulding, sand is left out of the mix, only cement and small-size gravel (pea-size) to deliberately make the rings porous. The three types of ring are illustrated in Figures 3.10.

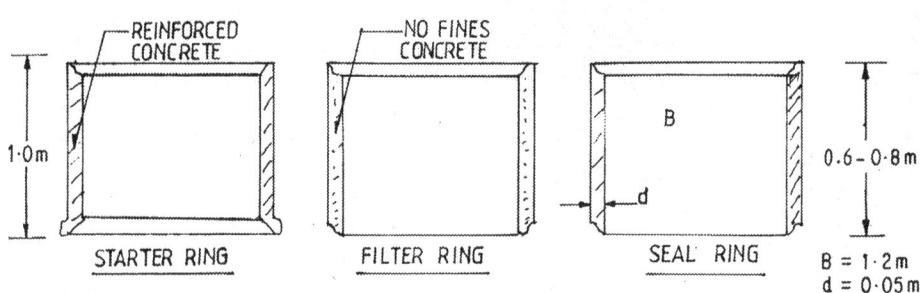

Figure 3.10: Three Types of Prefab Concrete Lining Rings

Nowadays, prefab concrete rings are readily available in markets close to most rural communities.

In some cases, a combination of prefab concrete rings and *in situ* concrete is used to reduce cost, *in situ* concrete is cheaper and is generally used in the upper sector of the well. This design makes it possible to deepen the well at a later time, either when the water table drops or water requirements increase, both making it necessary to increase the yield of the well. In this case, *in situ* cast concrete is then used to line the middle portion from which prefab rings descended. A cheaper design is the combination of prefab concrete rings with asbestos cement pipes. This however is not as amenable to future deepening as prefab/*in situ* concrete. Both are illustrated in Figure 3.11A and B.

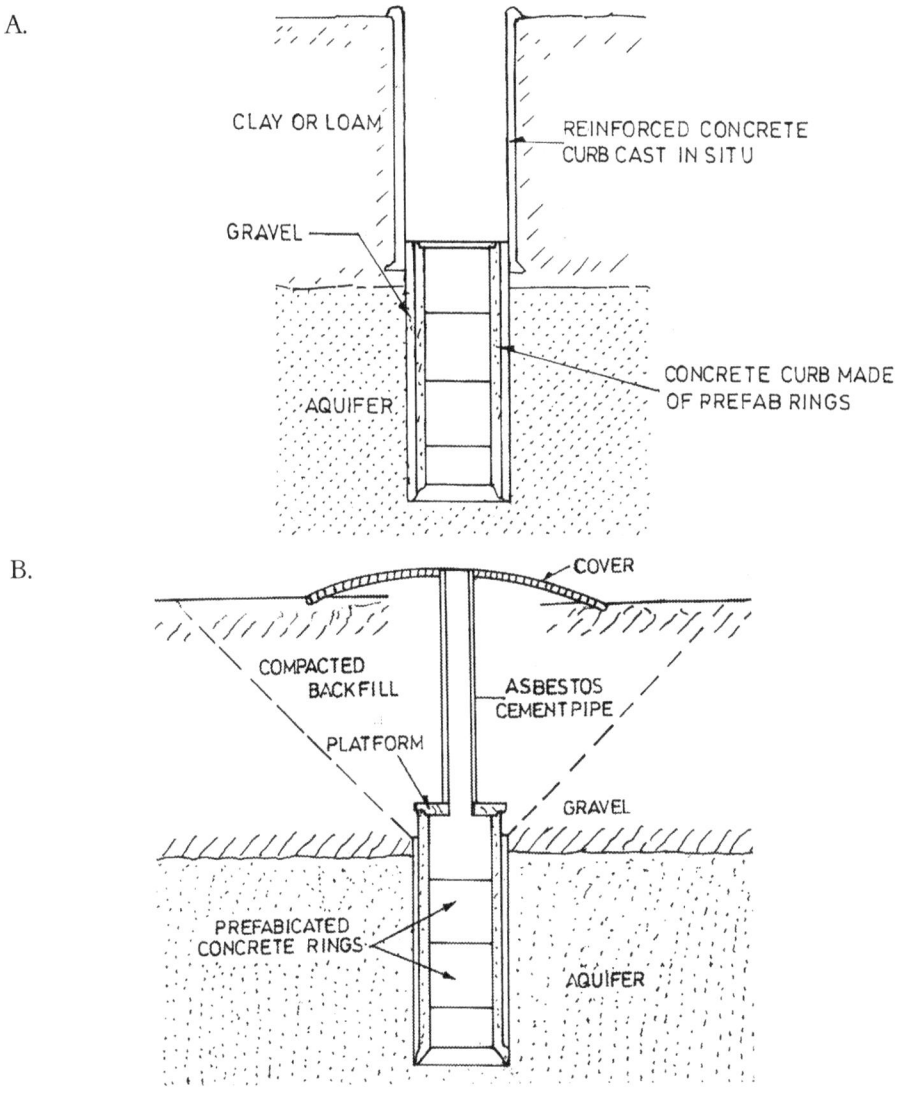

Figure 3.11A, B: Combination of Different Lining Materials

Graded gravel filters are sometimes laid at the bottom of the well to prevent sand from being drawn up. This is particularly helpful where a sub-artesian aquifer discharges into the well. Site conditions differ, but a normal graded gravel filter consists of 15 cm layers of 6 mm, 10 mm and 12 mm aggregates, with the smallest at the bottom. Each layer is levelled before the next is placed. A final capping of 20–25 mm gravel is used as a protection for the filter. Filter laying is done manually. Where overdrawing from the well is likely, either due to excess demand or poor yield, drawing from below the groundwater table may result in the movement of the aquifer particles, which may cause the well to cave in. An increased thickness of the bottom gravel filter will help prevent this overdrawing and the consequent collapse of the well.

The major limitation of dug wells is the maximum practical depth which can be reached by manual excavation: 60 m is considered to be this limit; deeper depths require bigger diameters for ease of sinking and pump installation and these additional requirements make digging beyond 60 m uneconomical. In addition, where hard rocks are encountered, they can be penetrated by only about 1 m, even with the use of explosives, which is also a very slow process. At this stage, boreholes become the obvious choice.

3.2.4.3 Location of Dug Wells

In rural areas and villages, the chances of polluting groundwater with bacteria and other pollutants (farm chemicals, fertiliser etc.) are very high. Wells must therefore be located as far as possible from likely sources of pollution.

Much research and studies have been carried out on this subject and the conclusion seems to be that it is difficult to make arbitrary rules. But as far as practicable, the following guidelines should be adhered to:

1. Well should be located uphill of latrines (privies) and septic tanks (cesspools). Where this is not possible, a distance of at least 15 m should be maintained between a well and a latrine as this distance is known to be able to prevent bacterial pollution of the well. In sandy soils, the distance could be as small as 7.5 m for the same reason. This requirement should be reconciled with the WHO recommendation that at least one well be available for every 200 inhabitants in rural areas and that users should not traverse more than 100 m to reach the source of water.

2. The ground surface immediately surrounding the well should slope away from it and should be well drained. In the case of public wells, a purpose-made drainage system should be provided for waste water and storm water.

3. Where a pump is installed, its platform or well cover as the case may be, should be at least 60 cm above the highest flood levels of nearby rivers; it should be generally protected against surface run-off.

4. When manholes are provided, the rims should project at least 8 cm over the surrounding surface and the manhole cover should overlap the rim. (See illustration in Figure 3.7).

5. When more than one well must be sunk and pumped simultaneously, any undue interference between the wells should be avoided. For this reason, especially if sunk into the same aquifer, the spacing should be 150–300 m or even more.

6. Before putting a completed well into use, and immediately after every repair, a well should be disinfected to neutralise any bacterial infection through workmen, equipment or surface water during construction/repair. The finished surfaces should be washed with a strong solution of chlorine, obtained by dissolving 50 g of chlorinated lime in 100 litres of water; i.e.

100 p.p.m. of available chlorine. The volume of water in the well should also be measured and an appropriate chlorine solution added to obtain an effective chlorine dose of 50–100 p.p.m. Application of the chlorine is usually made at different levels of the water, which is then agitated to ensure even distribution. The chlorinated water should be left in the well for 12 hours, after which it is pumped out completely.

3.2.4.4 Improvement of Existing Wells

Existing wells may require repairs and improvement, if found to be yielding unsafe water or dry altogether, due to a number of reasons. The problem of each such well should be investigated, paying particular attention to the following areas.

1. the location of the well in relation to nearby buildings and possible sources of pollution
2. the slope of the ground surface and the water table
3. the surface drainage around the well
4. the condition of the platform and well cover, if applicable
5. the condition of the lining and curb
6. the method of drawing water from the well
7. the presence of sediment in the well
8. whether or not there is the intrusion of surface water into the well
9. the water quality, especially its chemical and bacteriological quality.

In some cases, the condition of the well would most likely be such that it would be condemned altogether, especially due to location problems which cannot be solved. But in most cases, it could be possible to improve the well and make it safe and productive. The following is a list of steps that could be taken to improve wells, taking the peculiar problem(s) of each well into account.

1. Cleaning and deepening to improve capacity and physical property of water.
2. Installation of casing/lining, using an appropriate method, based on available materials (see Lining in Construction of Wells). If some form of lining already exists, it may be improved upon as in Figures 3.12A and B, or reconstructed as in Figures 3.12C.
3. Build a strong concrete or clay impervious platform (or apron) at least 1 m round and sloping away from the well.
4. The well casing (or lining) should extend at least 60 cm above the ground level
5. Provide an appropriate cover and provide a sanitary method of water withdrawal, preferably pump.

A.

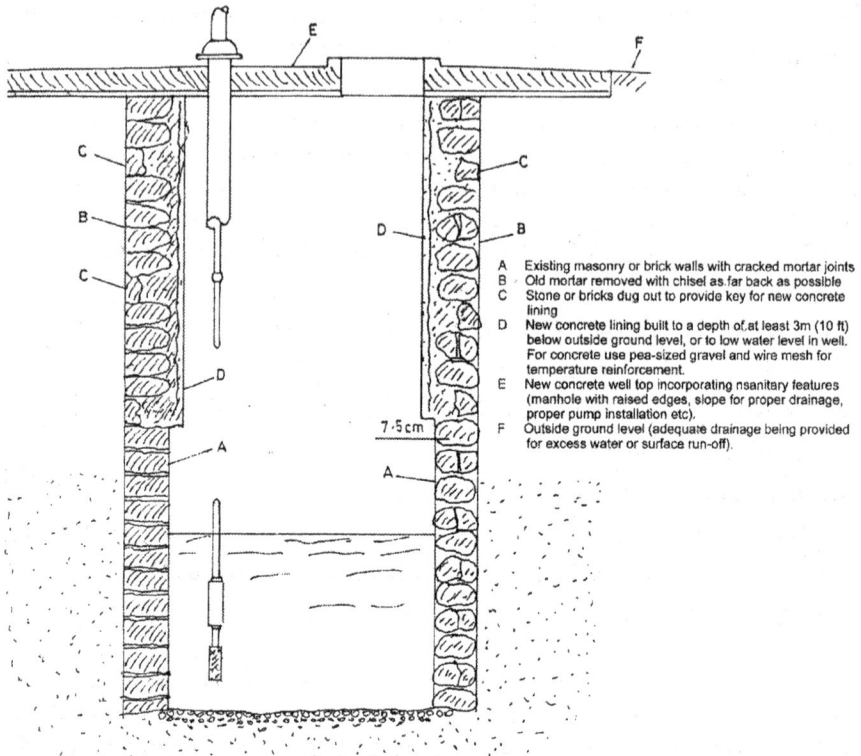

A Existing masonry or brick walls with cracked mortar joints
B Old mortar removed with chisel as far back as possible
C Stone or bricks dug out to provide key for new concrete lining
D New concrete lining built to a depth of at least 3m (10 ft) below outside ground level, or to low water level in well. For concrete use pea-sized gravel and wire mesh for temperature reinforcement.
E New concrete well top incorporating nsanitary features (manhole with raised edges, slope for proper drainage, proper pump installation etc).
F Outside ground level (adequate drainage being provided for excess water or surface run-off).

Figure 3.12A and B: Repair of Linings

B.

G. Blackfill with clay, well tamped in layers 15 cm (6 in) thick

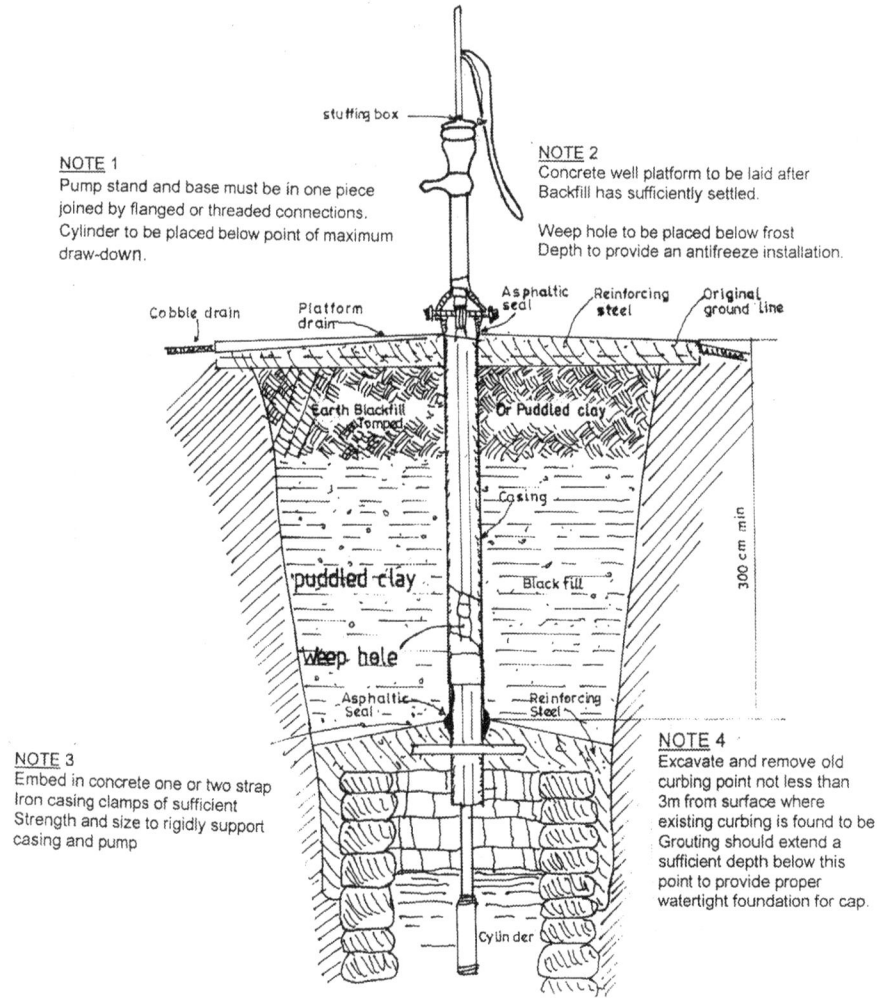

Figures 3.12C: Reconstructed Dug Well with Buried Slab

Where the problem is due or related to aquifer depletion, it could be solved by deepening the well, depending on whether or not it has reached the end of the aquifer; it could also be solved by reducing the pumping/withdrawal rate. In the former case, the solution would involve deepening the well and putting in appropriate gravel filter etc. In the latter case, pumping rate can be reduced by raising the suction pipe of the pump or introducing a thick graded gravel filter, described in Construction of Wells above, whichever is appropriate, depending on the withdrawal method in use. Where the problem has to do with damaged casing or screen, it is almost always advisable to replace them. Figure 3.13A and B, illustrate other methods of improving existing wells.

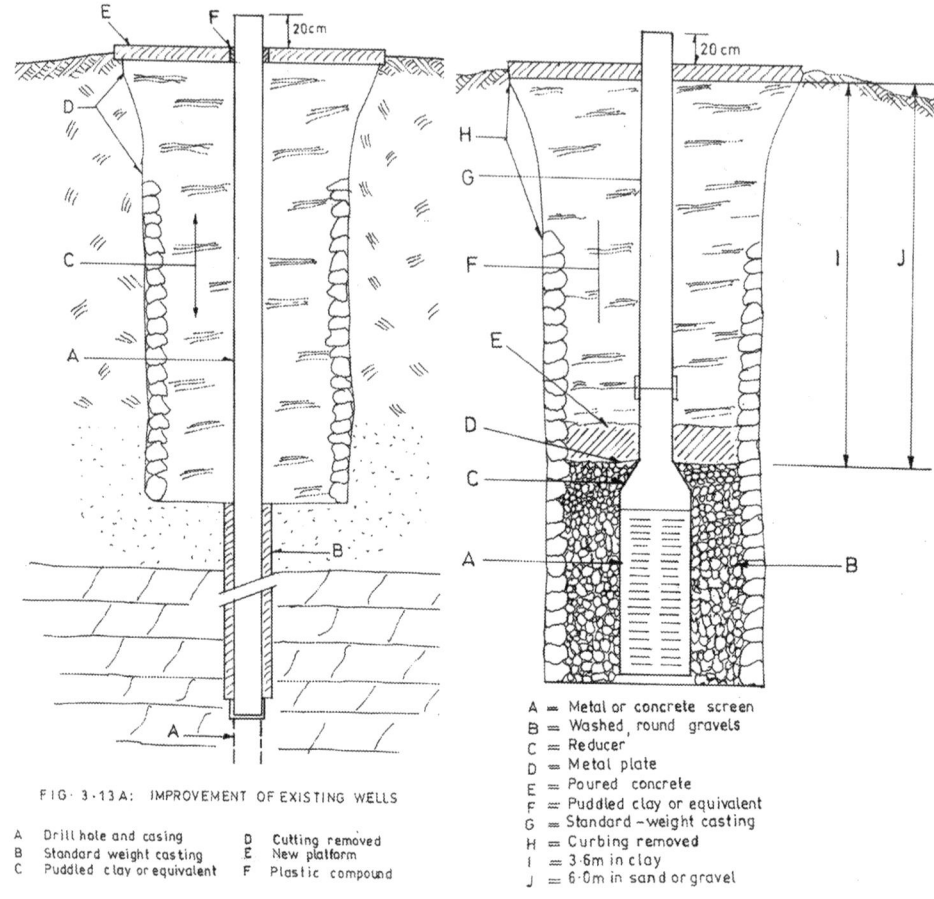

FIG. 3.13A: IMPROVEMENT OF EXISTING WELLS

A   Drill hole and casing
B   Standard weight casting
C   Puddled clay or equivalent
D   Cutting removed
E   New platform
F   Plastic compound

A = Metal or concrete screen
B = Washed, round gravels
C = Reducer
D = Metal plate
E = Poured concrete
F = Puddled clay or equivalent
G = Standard-weight casting
H = Curbing removed
I = 3·6m in clay
J = 6·0m in sand or gravel

Figures 3.13A & B: Other Methods of Improving Existing Wells

3.2.5 BOREHOLES

Boreholes differ from dug wells primarily from the depth at which water is extracted, their diameters, casing/lining and the delivery system which is a function of these other differences. They are usually 30–500 m deep, though they could be up to 1 km deep in arid regions.

Drilling boreholes is a specialised well construction technique, which will not be discussed in detail here. It suffices to know that once the aquifer is deeper than 6 m (but practically from about 30 m upwards), water can be extracted from it only by means of boreholes. Borehole drilling requires a Drilling Rig, equipped with a number of components and tools, capable of drilling in any geological formation to great depths; it requires considerable knowledge and experience, found only in specialised drillers, who should be contacted once a decision is taken to sink one.

There are two principal methods of construction used by drillers – percussion drilling and rotary drilling, applied in hard and soft formations respectively. Percussion drilling involves the alternate raising and dropping of the drilling tool – bit – in the borehole and a device for removing the loose excavated materials. Rotary drilling uses a rotating bit to cut or abrade the hole bottom. In both cases, devices are provided to remove the loose excavated materials, install screens and casings (linings) where necessary, develop (clean up) the borehole and install pump and accessories for water withdrawal.

The commonest pump types used these days are handpumps, especially in rural communities and for low-yielding wells (0.8–2 m^3/hr), and electric submersible centrifugal pumps for medium to high-yielding (2–4 m^3, medium; 4 m^3 and above, high) wells and more affluent communities. Borehole pumps are available in various sizes from about 50 mm diameter to supply heads of 100 m or more. A 100 mm diameter pump could supply about 4 m^3/hr, while one of 250 mm diameter could supply as much as 120 m^3/hr, depending on the head. Figure 3.14 illustrates a typical borehole water supply installation.

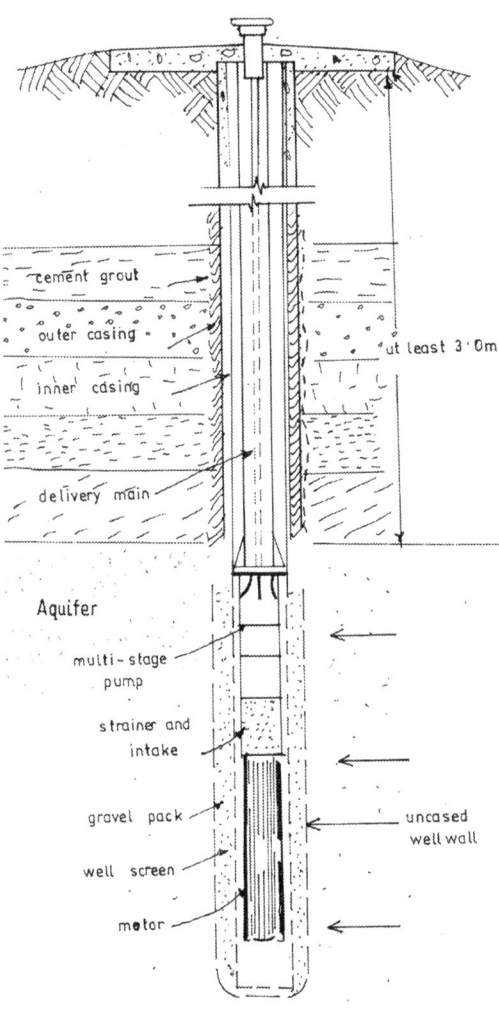

Figure 3.14: A Typical Borehole Installation

3.3 Groundwater Dams

3.3.1 INTRODUCTION

In the development of water resources, especially for rural communities, there is the need for a shift in orientation from large and expensive sophisticated schemes towards appropriate low-cost and socially acceptable schemes, adaptable to local techniques and conditions; groundwater dams seem to provide a suitable avenue for this shift.

Many developing countries, where groundwater dams are most suitable, are situated in regions of seasonal and highly erratic rainfall. This implies that water supply would involve storage during the rainy season for use during the dry season and in years of above average rainfall for use in those of below average. Where there are good aquifers within easy reach, using groundwater is a way out, though it too is sometimes depleted towards the end of the dry season. In other areas, either there are no aquifers at all or they are located so deep that only deep-drilled wells (boreholes) equipped with pumps are the means of tapping water from them. Most rural communities cannot afford this method.

This scenario brought groundwater dams into limelight in water resources development for rural communities, though it is not a new concept. It has been practised in several parts of the world – North Africa, Arizona, USA, and more recently in India, Brazil and Southern and Eastern Africa.

A groundwater dam intercepts groundwater flow and stores the water below the ground surface. Groundwater dams could also be used to divert groundwater flow to recharge adjacent aquifers or raise groundwater table in aquifers with low amounts of water.

There are two types of groundwater dams – Subsurface dams and sand-storage dams. Subsurface dams are constructed below the ground surface level and intercept the flow in a natural aquifer. On the other hand, sand-storage dams impound water in sediments caused to accumulate by the dam itself. The dam is built on the surface to accumulate sand and gravel, which ultimately stores water in the just-created artificial aquifer. (See Figures 3.15A and B).

Both subsurface and sand-storage dams are in response to specific human needs and geophysical circumstances, though in some cases, they can be combined if geophysical circumstances permit.

A: Sub-surface Dam

B. Sand-storage Dam

Figures 3.15A & B: General Principles of Ground Water Dams

3.3.2 ADVANTAGES OF GROUNDWATER DAMS

- They allow the use of water which would otherwise be wasted.
- By keeping water table close to the ground surface, they reduce irrigation needs because some water reaches growing crops by capillarity within the root zone.
- Because water is stored in a reservoir below ground level, evaporative loss is much lower than from an open water reservoir, i.e. tanks, dam reservoirs etc.

- Water economy is easier to achieve because of the use of buckets (and pumps), i.e. greater areas can be irrigated or more head of livestock can be watered in addition to domestic consumption.
- The breeding of pathogenic organisms like mosquitoes and snails is eliminated.
- No loss of land to the impounding reservoir as is the case in conventional dams.

One of the objectives of this handbook is to create the awareness in planners, agricultural and hydrological engineers and the general reader, particularly in developing countries, that groundwater dams can be used to harness water resources and to present the situations where they are applicable. It is also to guide the authorities and water resources engineers in particular, on the necessary studies, planning, design, construction and the necessary operation and maintenance (O&M) practices for sustained usage. The main focus is on small scale domestic water supply, livestock watering and small-scale irrigation as practised in Tanga in Zebilla District in the Upper East Region of Ghana, about 20 to 50 ha per location.

If properly sited and constructed, groundwater dams do serve their purposes. The following are the conditions to satisfy in groundwater dam siting and construction:

- Knowledge of the hydrogeological conditions of the site through proper investigation.
- Costs should be minimised – investigate with simple inexpensive methods, use extrapolations and inferences to reduce study/investigation costs.
- Construct in good weather, usually at the end of the dry season, just before the outset of the rainy season; use local materials as much as possible; follow basic engineering practices.

3.3.3 FACTORS WHICH FAVOUR GROUNDWATER DAMS

1. Climate

Arid areas with low, seasonal and irregular rainfall, where rainfall in the wet season needs to be stored for use in dry weather. These are the monsoon and dry tropical climate areas, where total annual rainfall is normally sufficient for domestic and agricultural purposes, but the seasonality of the rainfall creates a shortfall. Damming groundwater is therefore intended to bridge the dry season needs. Also, the monsoon rains sometimes fail and bring about serious consequences.

These areas of the world where precipitation (pptn) minus evaporation equals zero or negative (i.e. pptn. $- E_t = 0$ or is negative) constitute about 2/3 of the earth's surface. See also chapter I.

Figure 3.16: Areas of the World where Precipitation minus Evapotranspiration = 0 or Negative

These are the same areas for which the techniques/methods for developing water resources discussed in this book are recommended. Most of the currently documented groundwater dams are located in these regions.

TABLE 3.2   RAINFALL (R) AND POTENTIAL EVAPORATION (EP) DATA FROM SOME DRY AREAS

AREA	AVERAGE RAINFALL (MM/YR.)	POTENTIAL EP (MM/YR.)
Biskra, Algeria	180	1330
Tarfaya, Morocco	110	850
Moudjeria, Mauritania	170	1870
Lugh Ferrandi, Somalia	360	2060
North Turkana, Kenya	200–600	>2500
Machakos, Kenya	850	1600–1800
Dodoma, Tanzania	590	1110
Catuane, Mozambique	670	1300
Gross Barmen, Bamibia	400	2260
Palghat Gap, India	2000–3000	1550
Bartlett Dam, Arizona	350	3090
Nordeste, Brazil	500–1000	2000

SOURCE: ADAPTED FROM AKE NILSSON

In virtually all the locations listed above, potential evaporation is higher than average annual rainfall (R or Pptn − Et < 1). Damming groundwater will drastically reduce evaporation in these areas and so make more water available for use. It should be noted, however, that evaporation also occurs in groundwater, though considerably less for all categories of soil than open water evaporation. It has also been shown that the deeper the groundwater table, the lower the evaporation. It can be said fairly accurately that the finer the surface soil particles, the higher the evaporation because capillary movement of water is higher in fine-grained soils than in coarse-grained soils. These facts should be taken into account in designing groundwater and sand-storage dams.

2. Topography

Well-defined narrow valleys or river beds are preferable for groundwater dams – this reduces costs and makes it possible to estimate the storage volumes as well as control possible seepage losses. Storage volumes should be maximised while dams are kept as small as possible to keep costs low.

Groundwater dams are built primarily to stop the depletion of groundwater through natural flow, which is a function of the gradient of groundwater table which is itself, a function of the topographic gradient. The range of good longitudinal gradients for groundwater and sand-storage dams is 0.2 to 4%, though in extreme cases groundwater dams have been constructed on gradients of up to 16% (Ake Nilsson).

Balancing between longitudinal and lateral movement of sediments along valleys and river courses, the most favourable sites for groundwater dams are generally found on the gentle slopes in the transition zone between hills and plains. A good dam site, based on optimal topographical conditions, is at a narrow passage between outcrops through which the groundwater is drained from an aquifer of even thickness in a wide valley on a gentle slope (Figure 3.17).

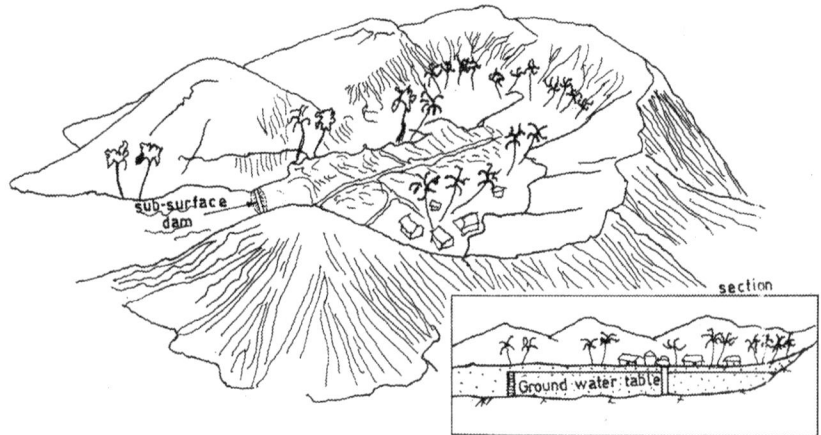

Figure 3.17: Topographical Features of a Good Groundwater Dam Site

3. Hydrogeology

The most suitable aquifers for groundwater dams are river beds made up of sand or gravel; others are *in situ* (residual soils) weathered layers and deeper alluvial formations, though not as good as the former. In this type of aquifer, average specific yields of about 25% could be obtained from the former whereas it is as low as 7.5% in the latter, depending on the proportion of fine materials. A balance must be struck between porosity and hydraulic conductivity (permeability) of aquifers; care must also be taken to avoid fractured zones through which the stored water could be lost through seepage.

Most groundwater dams are in unconfined aquifers and so are generally from 6 to 9 m deep. However, a groundwater dam was reported to have been built across a confined aquifer at a depth of about 25 m on the island of Kaba in Japan. Such dams are highly sophisticated, expensive and outside the range of possibilities in developing countries.

Groundwater reservoirs are generally recharged by lateral groundwater flow, but when the stored water is used for irrigation upstream of the dam (as in the Upper East Region of Ghana), there will also be some recharge from return flow, provided the surface soil is permeable.

Groundwater dams are generally founded on solid bedrock; to avoid seepage; therefore, the dam should be tight because it is difficult to stop seepage through fractures, even if detected during dam construction. Other impermeable formations could also serve as reservoir bottom, e.g. alluvial or uppermost part of a weathered layer, especially if sufficiently thick.

4. Sediments

Accumulation of sand-gravel sediments upstream of a sand-storage dam is the objective of a sand-storage dam. This accumulation is the final result of a series of activities involving detachment of soil particles, their transportation and deposition under the factors of rainfall, slope, particle size and land use. A catchment characterised by steep slopes, scanty or no vegetative cover, is more favourable than gentle slopes with thick vegetation! This is contrary to the views of agricultural and hydraulic engineers!

3.3.4   USER ASPECTS OF GROUNDWATER DAMS

1. Water stored is generally for irrigation and domestic purposes i.e. human and livestock consumption; it is rarely used for industrial purposes (see Table 3.3).
2. In terms of water usage, there is a basic functional difference between subsurface dams and sand-storage dams; the former are almost 50:50 for irrigation and domestic use, whereas almost 90% of sand-storage dams are for domestic purposes.

TABLE 3.3   PURPOSES FOR WHICH GROUNDWATER DAMS ARE BUILT.

MAIN WATER USE	SUBSURFACE DAMS		SAND-STORAGE DAMS	
	NO OF SCHEMES	APPROX. VOL. (M^3)	NO OF SCHEMES	APPROX. VOL. (M^3)
Irrigation	9	13,000–10^6	1	6,000
Drinking Water (domestic or cattle)	7	400–2,000	9	50–12,000

SOURCE: AKE NILSSON

3. Water extraction methods depend on the intended use of the water. For irrigation, wells and motorised pumps can be economical, whereas gravity or small handpumps will more often than not be the choice for domestics (see Figure 3.18).

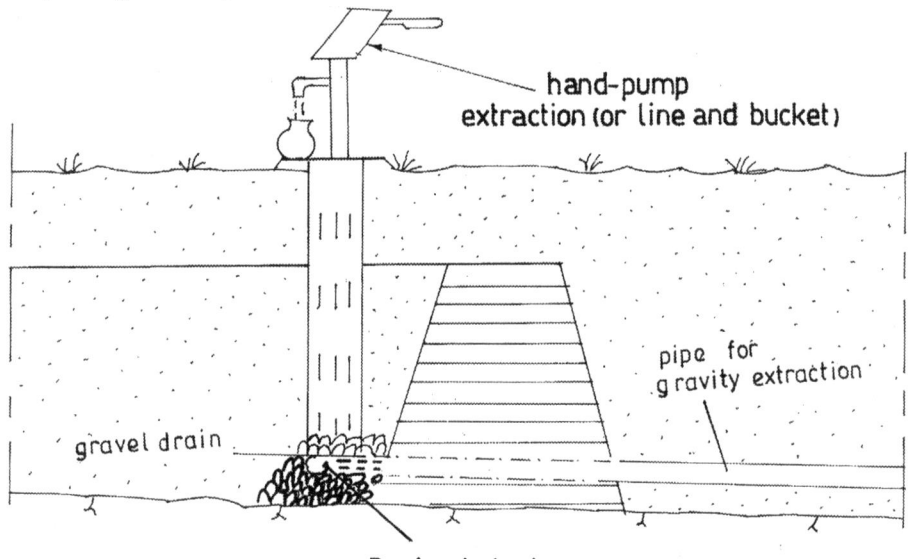

Figure 3.18: Water extraction method as a function of use

A summary of favourable characteristics of suitable sites is as follows:
1. Scarce seasonal and erratic rainfall
2. Gentle gradients (terrain)
3. Thick and well-defined river beds, made up of coarse sediments and underlain by low-permeable layers
4. Availability of dam construction materials
5. Users and planners must be linked early in the planning stage, the awareness must be created in would-be users and a focus on the benefits will help to encourage development efforts.

3.3.5 ECONOMIC FACTORS

Water volumes stored in groundwater dams are generally small, hence, to achieve an acceptable benefit:cost ratio, costs must be kept low; to this end, avoid the use of expensive investigation methods and construction materials. Furthermore, benefit:cost analysis of this kind of scheme is not easy – present and future benefits are difficult to evaluate/estimate; what they are and their worth in economic terms are difficult to decide. While it is possible for schemes, which supply water for irrigation, it is not so for water supply schemes; domestic production and health could be positive effects, but how can they be defined economically? More so when water supply alone does not improve people's health, but other inputs like sanitation, health education etc., also play roles. The result of any such analysis therefore can at best give only an indicative figure of the order of magnitude of the present and future benefits – they should be used with care. (More details in chapter X – Economics of Water Resources Development).

3.3.6 COST OF WATER FOR DIFFERENT MATERIALS

1. For sand-storage dams constructed in India, Kenya, Botswana and Namibia between 1956 and 1984, the costs per m^3 of estimated storage were US$ 0.1 for brick and a combination of concrete and stone masonry, whereas it was between US $1.0 and 5.0 for concrete dams – (see Table 3.4).

TABLE 3.4   COST OF STORED WATER IN SELECTED SCHEMES

TYPE OF DAM	COUNTRY	COST/M^3 OF ESTIMATED STORAGE (US$)	YEAR OF CONST-RUCTION	REFERENCE
Sand-storage dam/concrete	Botswana	2–5	1984	UNESCO, 1984
Sand-storage dam/concrete	Kenya	1.3	1984	UNESCO, 1984
Subsurface dam/brick	India	0.1	1963	Destouni and Johansson 1987
Sand-storage dam/concrete and stone masonry	Namibia	0.1	1956–1965	Stengel, 1968

2. Groundwater and sand-storage dams also yield more water than the conventional surface water dams, mainly due to the effect of evaporation. Quoting Burger and Beaumont, 1970, Ake Nilsson reported that while a conventional dam, 12 m high and 80 m long was found to yield only 20% of the impounded water, i.e. 46,000 m³/year, a sand-storage dam of corresponding size yielded 70% of the water stored in the sand, i.e. 41,000 m³/year. Though the sand-storage dam cost slightly more because it had to be constructed over several years, eight, in this case, the fact that water from the conventional dam has to be treated to qualify for domestic and drinking purposes makes it more expensive at the end of the day. The surface dam reservoir is also subject to siltation and therefore has a limited life span. Table 3.5 shows the Cost Analysis of this example.

TABLE 3.5 CASH-FLOW ANALYSIS FOR SAND-STORAGE AND CONVENTIONAL WATER SUPPLY PROJECTS

A.

Type of project	Year 0	Year 1	Year 2	Year 3	Year 4
1. SAND-STORAGE PROJECT					
a. Capital expenditure	96,500	60,000	-	21,000	-
b. Yield in m³/year	391,000	-	-	-	1,000
2. CONVENTIONAL DAM					
a. Capital expenditure	89,600	95,000	-	-	-
b. Purification ac/m³	13,200	-	1,740	1,530	1,340
c. Losses 0.2 c/m³	4,400	-	600	580	540
Total a, b, & c.	107,200	95,000	2,340	2,110	1,880
d. Yield in m³/year	330,000	-	46,000	43,000	40,000

B.

Type of project	Year 5	Year 6	Year 7	Year 8	Year 9
1. SAND-STORAGE PROJECT					
a. Capital expenditure	18,000	-	-	14,000	-
b. Yield in m³/year	3,000	6,000	9,000	13,000	20,000
2. CONVENTIONAL DAM					
a. Capital expenditure	-	-	-	-	-
b. Purification ac/m³	1,150	1,000	890	770	680
c. Losses 0.2 c/m³	510	480	440	410	380
Total a, b & c.	1,660	1,480	1,330	1,180	1,060
d. Yield in m³/year	36,500	34,000	31,500	29,000	27,000

C.

Type of project	Year 10	Year 11	Year 12	Years 13-30	Years 31–
1. SAND-STORAGE PROJECT					
a. Capital expenditure	-	-	-	-	-
b. Yield in m³/year	27,000	37,000	41,000	41,000	41,000
2. CONVENTIONAL DAM					
a. Capital expenditure	-	-	-	-	-
b. Purification ac/m³	592	514	453	390-0	-
c. Losses 0.2 c/m³	350	330	300	270-0	-
Total a, b & c.	942	844	753	660-0	-
d. Yield in m³/year	25,000	23,000	21,500	19,500-0	-

SOURCE: AKE NILSSON

This table shows the following:
1. The total capital cost of the sand-storage dam, built over an 8-year period is $209,500
2. It started with a yield of 1,000 m³ in year 4 and continued steadily, reaching its maximum of 41,000 m³ in year 12 of the project; it maintained this maximum yield for 19 years thereafter.
3. On the other hand, the capital cost of the conventional dam, built over a 2-year period is $184,600.
4. The conventional dam yielded its maximum of 46,000 m³ in year 2 and declined steadily thereafter, due to evaporation and siltation.

3.3.7 PLANNING AND INVESTIGATION METHODS

Groundwater dams are useful for water conservation and supply under certain specific conditions, which are not universally applicable elsewhere and therefore cannot be extrapolated. Every proposed site should be investigated and other alternatives considered before a final decision is taken. For example:

- the nature and extent of the bedrock or other underlying impervious layer should be ascertained to ensure that it can hold water – seepage at that depth cannot be corrected once the dam is in place
- the nature of the aquifer must be ascertained to be able to estimate the storage volume and hence the expected yield.

3.3.8 INVESTIGATION METHODS

1. Topographical, geological and thematic maps and satellite imagery are the easiest methods to use; they give information on geology, climate, hydrology, hydrogeology, soil, vegetation and land-use.

2. Reports from previous studies – these could give information on erosion and sedimentation
3. Air-photo interpretation is the best method (instrument) of finding groundwater dam sites, especially when combined with map studies and ground control.
4. Where stream flow records are lacking, peak flow level could be estimated from marks left on objects, which usually give a good indication, local people could also be interviewed to give indicative answers on flood and sediment accumulation.

3.3.9   DESIGN AND CONSTRUCTION OF GROUNDWATER DAMS

Subsurface Dams

Design Notes:
a. Storage volumes range from a few 100 m^3 to several millions; design therefore differs, depending on the storage capacity. In this book, we deal only with small-scale dams of height 3–6 m. The basic design is as follows:
   i. excavate a trench across a valley or stream bed to the bedrock or other impermeable stratum
   ii. build a dam in the trench, on a concrete footing and backfill with the excavated material,
b. Because of the gravelly and or sandy nature of the material in which excavation is made, a trench side slope of 3:1–3.5:1 is generally adopted to keep the trench stable.
c. Because depths are generally 3–6 m, earth work is manually done. In this case, it is easier to control the slope by excavating in steps with horizontal levels every 1.0 to 1.5 m, depending on the stability of the soil being excavated, though this will increase the overall width of excavation (see Figure 3.19).
   i. Groundwater dams are usually built at the end of the dry season when there is very little water left in the aquifer; even then, it may be found necessary to pump out any water accumulating in the trench to create a dry environment for dam construction.
   ii. The dam (impermeable screen, dike etc) is then built on concrete footing.
   iii. After construction, the trench is backfilled with the excavated material and properly compacted manually with watering to achieve maximum compaction – failure to do this may cause dam failure.

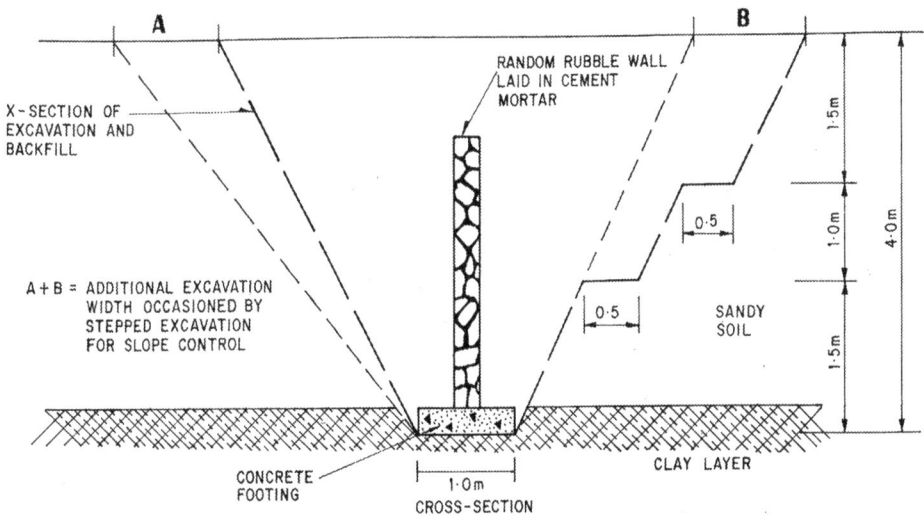

Figure 3.19: Cross-section of ground water dam excavation

Characteristics of dams built with different materials

Subsurface dams are generally built of clay, concrete and stone masonry as illustrated in Figures 3.20A, B and C.

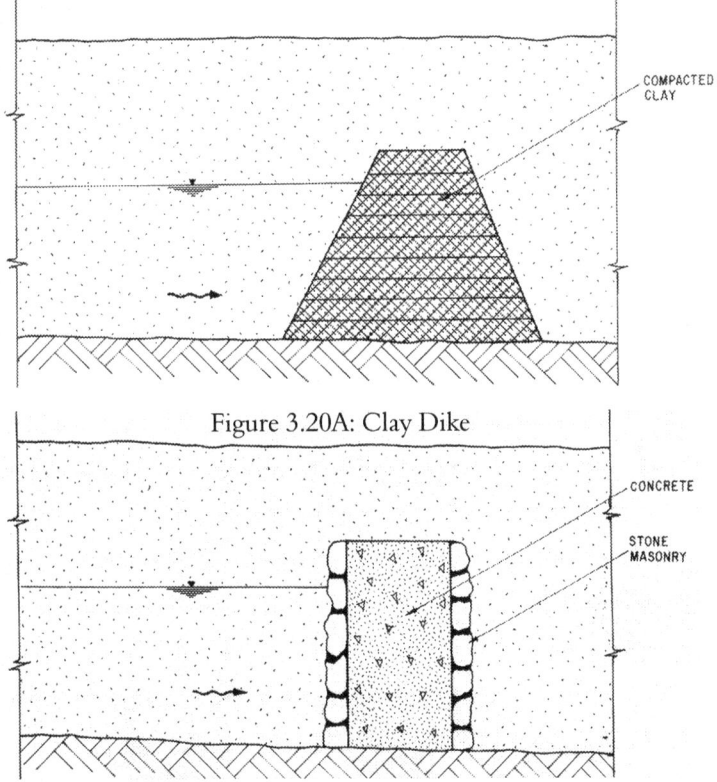

Figure 3.20A: Clay Dike

Figure 3.20B: Concrete Dam

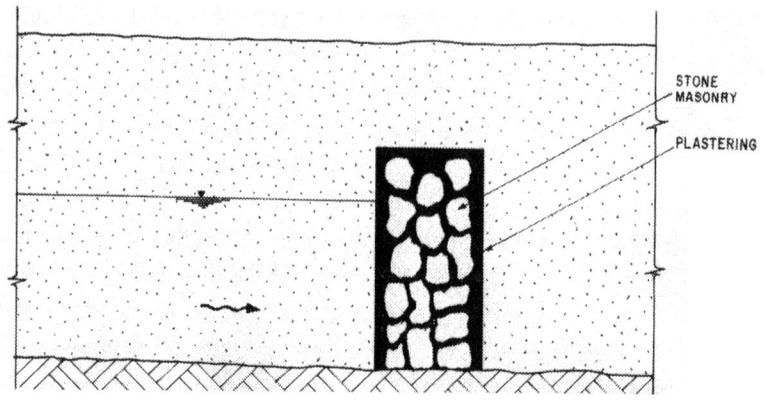

Figure 3.20C: Stone Masonry Dam

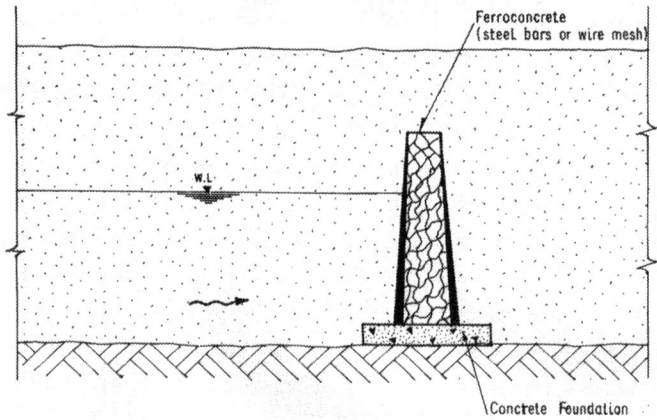

Figure 3.20D: Ferro-concrete dam

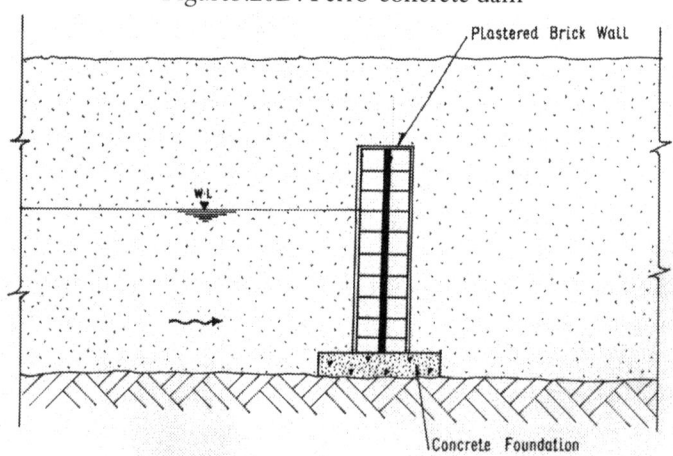

Figure 3.20E: Brick Wall Dam

Figures 3.20A–E: Groundwater Dams Built with Different Materials

Dams built with these materials usually range in height from 3 to 6 m.
1. Clay dam built in thin compacted layers, one after the other in the order indicated. It is suitable for small schemes in highly permeable aquifers like sandy river beds of limited depth. The likely danger of erosion of the dam can be averted by covering the dam with polythene sheets before backfilling. Similarly, in the event that the aquifer is pumped dry in the dry season, the clay dam may crack; to avoid this, the dam should be of 2–3m thick to keep it sufficiently moist throughout the dry season.
2. Concrete dam protected with stone masonry. It requires more advanced engineering practice, which calls for the use of skilled labour.
3. Dam built purely of stone masonry. It is particularly suitable for places where stone is available locally; it also requires skilled labour.
4. Ferro-concrete dam – it requires the use of form work, though it has the advantage of requiring very little material to achieve a strong wall (dam).
5. Brick wall, plastered to make it watertight – this also requires skilled labour.

The design chosen should be based on the local availability of materials. The crest of the dam is usually about 1 m below the ground surface to:
   a. avoid water logging in the upstream area and
   b. avoid erosion damage to the dam if exposed to the surface.

3.3.10 WATER EXTRACTION FROM GROUNDWATER DAMS

A drain is usually incorporated into the dam along its upstream base. The drain consists of gravel or a slitted pipe surrounded by gravel filter. The function of the drain is to collect and lead the water to a well or gravity delivery pipe downstream.

The well for extraction may be placed in the reservoir or in the river bank for erosion control purposes. Gravity extraction is suitable if the community to be served is located downstream of the dam site, given suitable topographic conditions. Where this is possible, extraction is not only cheap but also avoids the expenses of pump installation and operation and maintenance (O&M). Both options are illustrated in Figure 3.18.

For irrigation purposes as in the Andra Pradesh area of South India and in the Tanga and Namoo districts of the Upper East region of Ghana, large diameter wells are constructed all over the area – 20 ha in Ghana – fitted with handpumps or motorised pumps for the irrigation of choice crops like vegetables – onions, tomatoes and leafy vegetables.

Sand-storage Dams

The flow of water in the stream/valley to be dammed will determine the design of the dam as regards stability and sedimentation. The important thing to keep in mind is that the dam has to withstand the maximum peak flow in the stream, which must also be discharged without eroding the banks.

With respect to sedimentation, the basic idea is to limit the height of each stage so that the flow velocity is kept high enough for fine particles to pass while coarse particles settle. The construction principles of a sand-storage dam are illustrated in Figure 3.21. The typical height of sand-storage dams is 1–4 metres. The height of each stage is generally determined from experience, usually 1–1.5 m. Notice that the fact that the dam is built stage by stage over a period of time increases the overall cost. This could be avoided by building the whole dam at once and leaving a notch in the middle, which allows excess flows to pass with fine particles. The notch is then filled in subsequent season(s). This type is illustrated in Figure 3.22 from a dam in Machakos, Kenya.

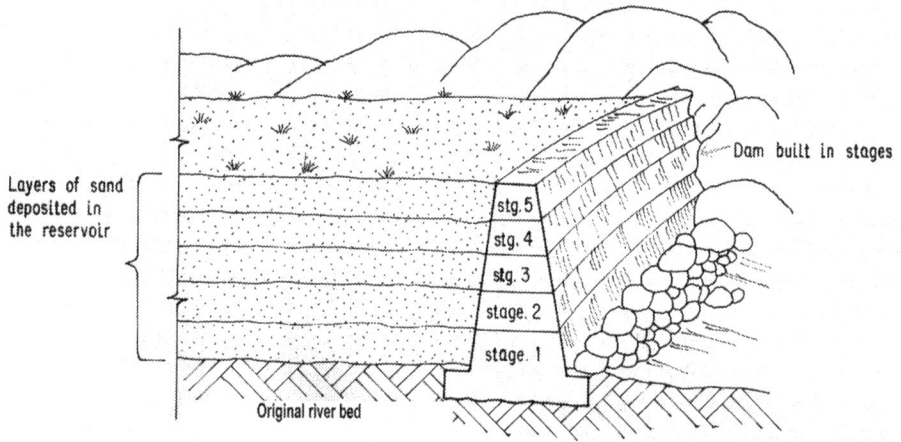

Figure 3.21: Construction Principle of Sand-storage Dam

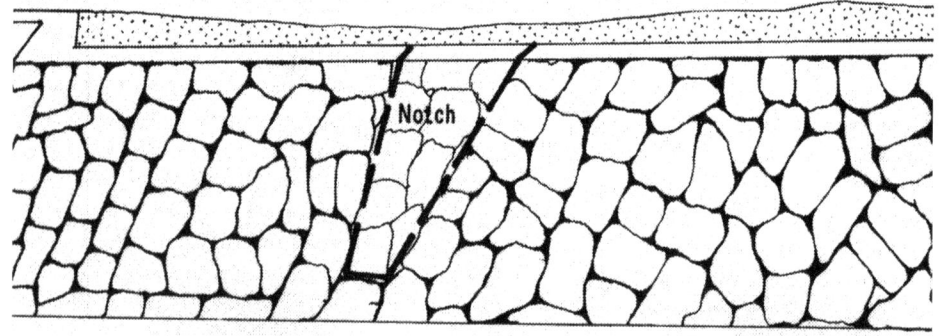

Figure 3.22: Filled-in notch, Kyandili sand-storage dam, Machakos, Kenya

3.3.11 DAM CONSTRUCTION

Figures 3.23A–F illustrate the different types of sand-storage dams, the concrete and masonry ones being the commonest. They fulfil the basic requirement of sand-storage dams, namely that:

1. They must be sufficiently massive to withstand the pressure from the sand (aquifer) and water stored in it, and
2. They must be watertight.

Where the dam is not absolutely watertight, as in Figure 3.23F, the water seeping through the dam (which must not be erosive) can be collected in troughs for watering cattle downstream.

Sand-storage dams must be protected against river-bank erosion. This can be avoided by projecting the dam several metres into the two banks or building wing walls of sufficient dimensions at the bank ends of the dam. The dam must also be protected at its toe by aprons of appropriate dimensions.

Spillways of appropriate dimensions also must be provided to discharge the overflows during peak flows.

A: Concrete Dam

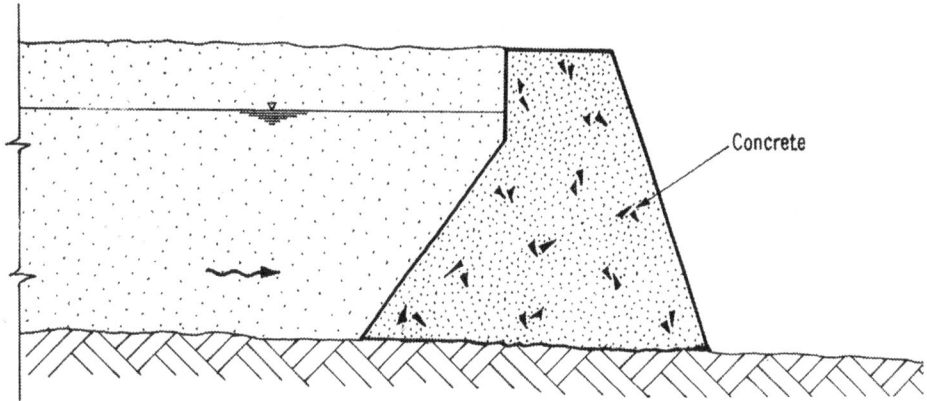

B: Stone Masonry dam

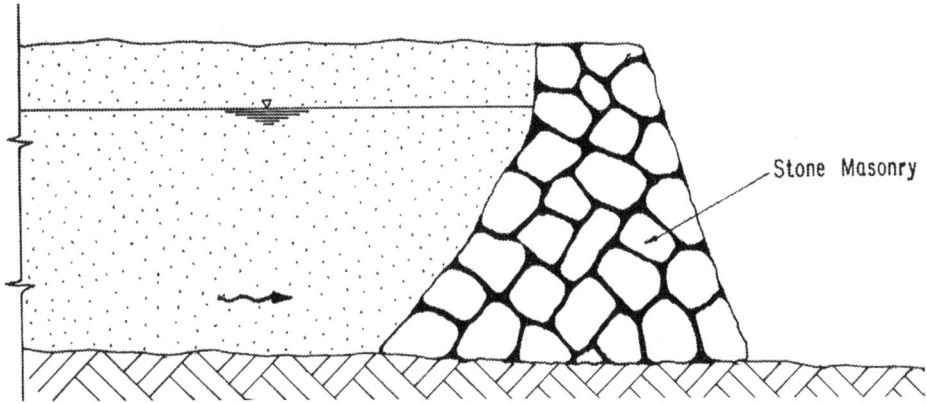

C: Gabion with Clay Cover

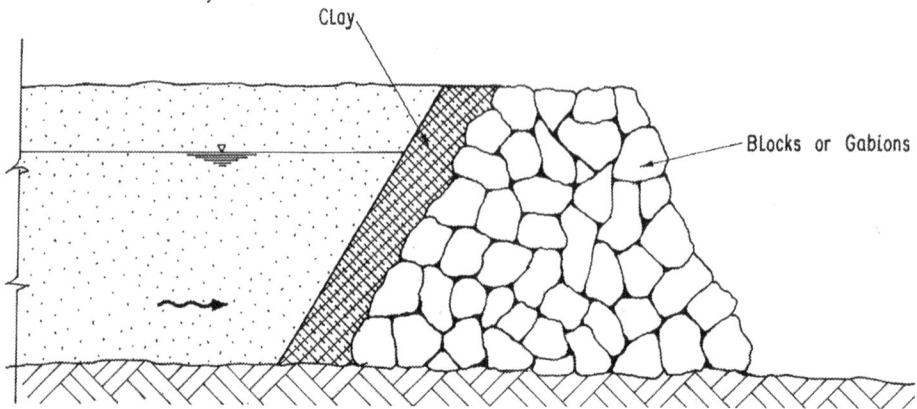

D: Gabion with Clay Core

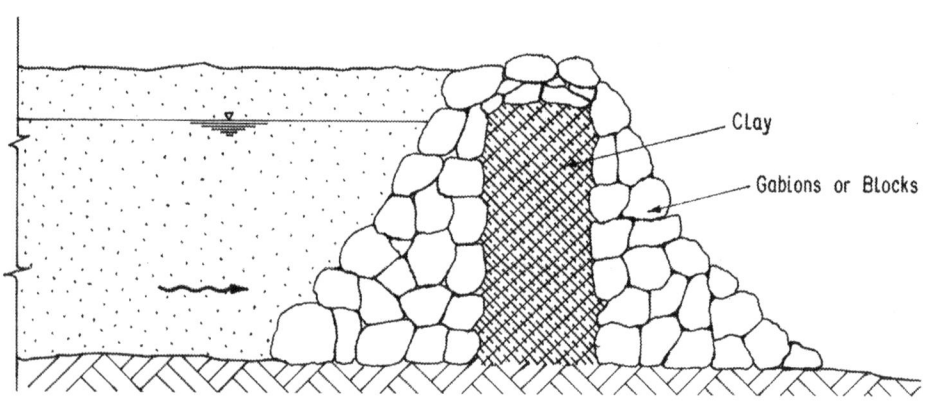

E: Stone-fill Concrete

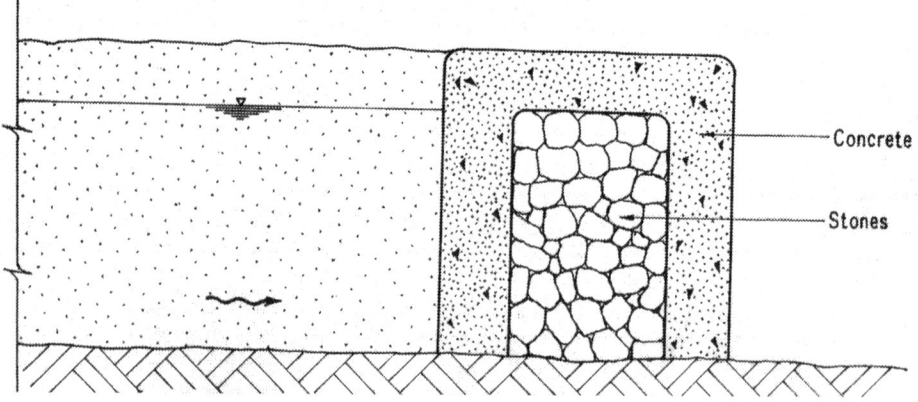

F: Stone Dam (Not watertight)

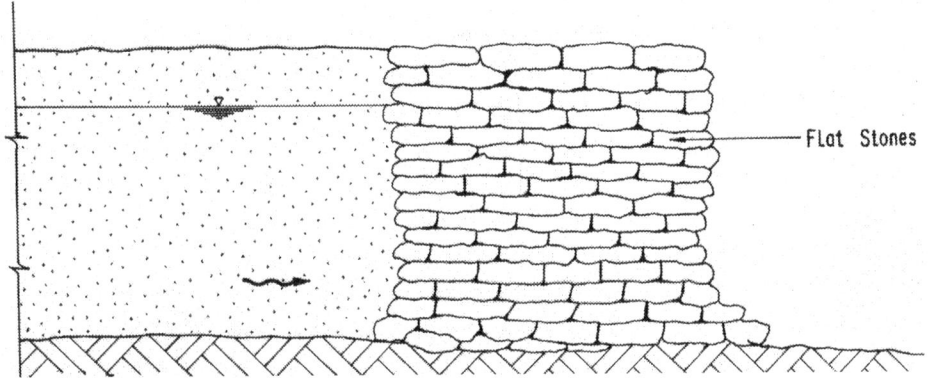

Figure 3.23: Sand-storage Dams

Water extraction from a sand-storage dam is basically the same as for a sub-surface dam, Figure 3.24 illustrates extraction alternatives.

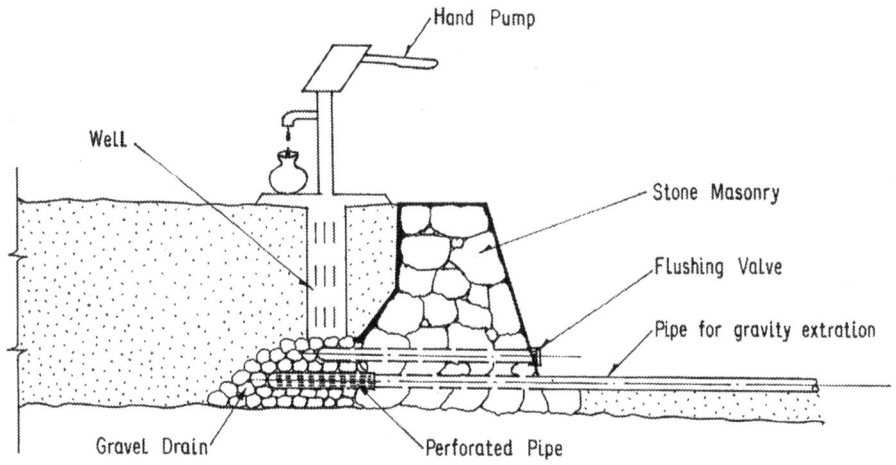

Figure 3.24: Extraction Alternatives from Groundwater Dams

Environmental Impact

If properly planned and implemented, a groundwater dam scheme should not have a negative impact on the environment.

Intercepting the natural groundwater flow, however, may have some effect on the groundwater conditions downstream. Such possible effects must be thoroughly considered at the planning stage.

The risk of possible waterlogging of upstream areas is negated by keeping the dam crest below ground level or by having a sluice in the dam.

The fact that the groundwater will rise introduces the risk of pollution. If the water is used for drinking, it may be necessary to fence or grass the recharge areas.

Where the water stored by a subsurface dam is used for irrigation, the return flow may recharge the aquifer. But this may lead to salinisation of the groundwater, especially if the following seasonal rains are not sufficient to wash out the salts. There may also be some salinisation of the soil and groundwater due to evaporation, if the groundwater table is shallow.

These and other possible impacts should be thoroughly investigated at the planning stage and conscious design considerations incorporated to minimise negative impacts. This subject is covered in greater detail in chapter IX.

3.4   Utilisation of High Groundwater Table (Dry Farming)

This is the cultivation of low-lying areas with high groundwater table, usually old river courses and flood plains of frequently flooded rivers, mainly for the production of vegetables. The water table is high enough to supply sufficient moisture to the root zone of growing crops by capillary movement of water.

Dry farming, as it is called, has been practised from time immemorial and has been the source of vegetables, seedlings of citrus and other tree crops, flowers and even cereals, particularly maize, for ages.

Though seasonal, it has been the source of livelihood for many communities because the production of these crops in the dry season, puts a high premium on them such that they command a reasonably high market price to justify the farmers' efforts. It is seasonal because the cultivated land is generally flooded during the rainy season, or the water table too high to be cultivated; it is known that some farmers grow paddy rice on some of these parcels of land, but the holdings are usually too small for such rice cultivation to command attention.

The practice is very well known to the inhabitants of the low-lying areas – both the agronomy and the field preparation – and does not require any particularly skilled input. It is mentioned here as one of the sources of water for agricultural production.

Chapter IV
SURFACE WATER

4.1 Introduction

Surface water derives essentially from rainfall and is the main source of water for agricultural purposes. It can be harnessed from the following sources:

i. utilisation of wetlands for crop production
ii. diversion from rivers and lakes
iii. interception/storage of overland flow
   - ponds and valley tanks/dug-outs
   - dams

4.2 Utilisation of Wetlands

Wetlands in this case refers to flood plains, freshwater marshes and swamps. They are good sources of water for human, livestock and crop production, as well as aquaculture. Their uses can be categorised as follows:

1. Use for crop production without being drained i.e. cultivation of crops that are adapted to waterlogged conditions as in the cultivation of rice and aquatic fodder (water ipomoea). This use has minimal disturbance on the ecological and hydrological regimes.
2. Construction of fish ponds for fish farming and duck farming.
3. Construction of surface ponds for livestock watering and provision of water for human domestic use, which has the added advantage of draining the surface soil for the cultivation of certain species of rice, finger millet, sorghum, local beans and sweet potatoes. This is the case in Kisii area in Western Kenya. This is called low intensity drainage.
4. Direct utilisation as dry season grazing land for livestock.

All these practices require very little skilled knowledge and when properly articulated, the practice results in a sustainable utilisation and conservation of wetlands.

4.3 Diversion/Pumping from Rivers and Lakes

In very rare cases, diversion from a river or lake may be one of the possible options of making water available for irrigated agriculture (as in the Ahmadu

Bello University's College of Agriculture, Bakura, Sokoto State Nigeria, where water is pumped from Lake Natu into a service reservoir over a distance of some 2 km, to supply the school farm, where students are trained in irrigation engineering and agronomy). In the event that it becomes important to consider diverting water from rivers and lakes, the following points should be borne in mind:

1. Wherever possible, river or lake intake should be located upstream of inhabited areas to reduce pollution and encourage gravity delivery.
2. The intake should be placed well below the water surface, to take up cooler water; it should also not be less than 1 m above the river/lake bed to avoid sediment and suspended matter being moved into it.
3. It should be located some distance from the shore – to prevent bank erosion, it should also be large enough to minimise entrance velocities which should be about 15 cm/sec. (0.15 m/sec.)
4. It could require a submerged dam/weir downstream across the river to provide a constant sufficient depth above the intake pipe or channel and for the settling of suspended matter (to reduce turbidity) and for keeping floating leaves etc from obstructing the intake structure.
5. A floating intake could be considered if the circumstances allow, e.g.
   - depth of water in the river/lake
   - physical location and
   - degree of permanency

   The floating intake could be made of empty oil drums, held in place by a frame, supporting an inlet hose.
6. Intakes should be designed to function with minimum attendance.

The following figures illustrate annotated typical river and lake intake structures/constructions

i. where the river does not have a load of boulders or rolling stones that could damage the intake, a simple unprotected intake is generally adequate (see Figure 4.1A)

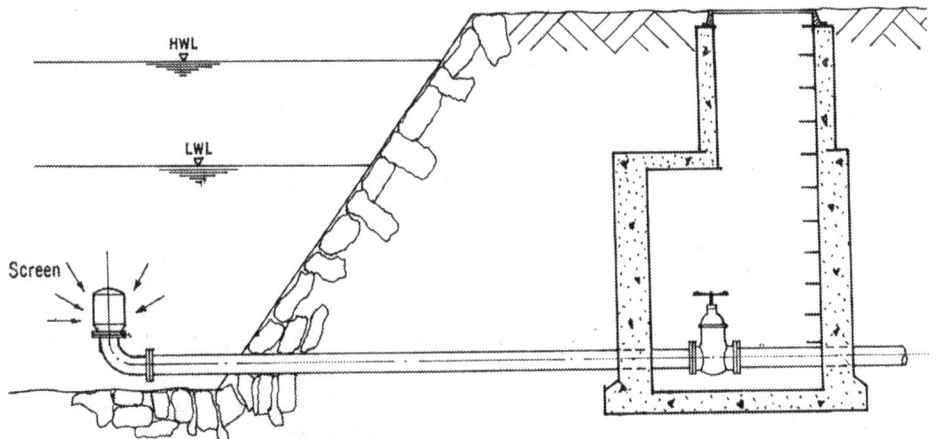

Figure 4.1A: A Simple Unprotected River Intake

ii. Where it is necessary to protect the intake against boulders and rolling stones, a protected intake as in Figure 4.1B may be suitable. Notice that a baffle may be necessary to keep debris and floating matter like tree trunks/branches out of the intake.

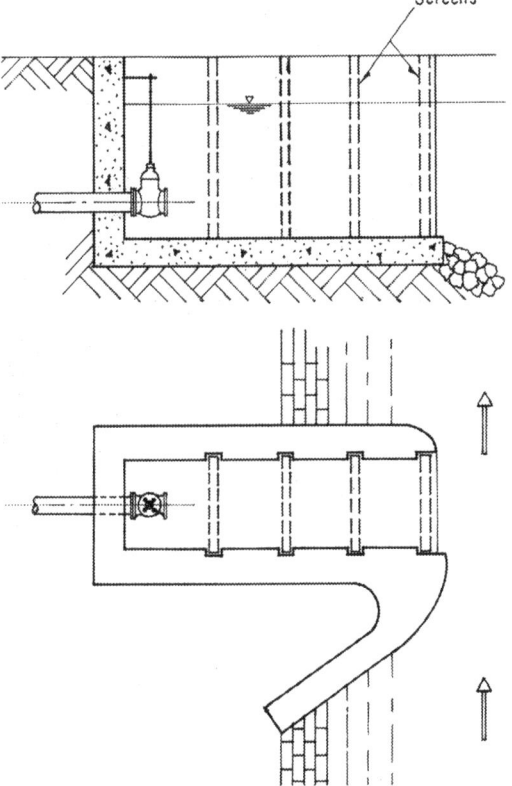

Figure 4.1B: A Protected River Intake

iii. Generally, pumping is necessary in the case of diversion from river sources. Where the level difference between the high and low water levels does not exceed 3.5–4 m, a suction pump positioned on the river bank is used (see Figure 4.1C). But if the level difference is more than 4 m, a more elaborate design is necessary (Figure 4.1D). In this example the river water is collected with an infiltration drain gallery laid under the river bed. Under gravity, the water flows into a sump, from where the water is pumped out with a submersible pump, since a suction pump placed above the ground may not be effective at such a depth.

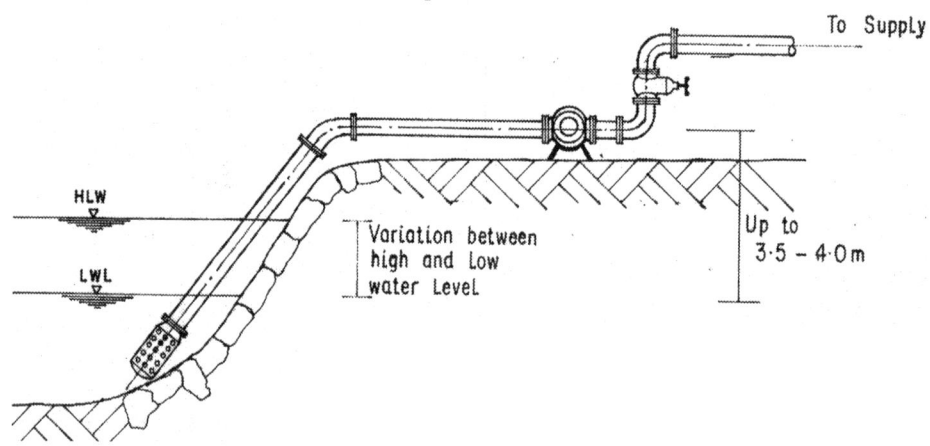

Figure 4.1C: A Suction Pump Positioned on the River Bank

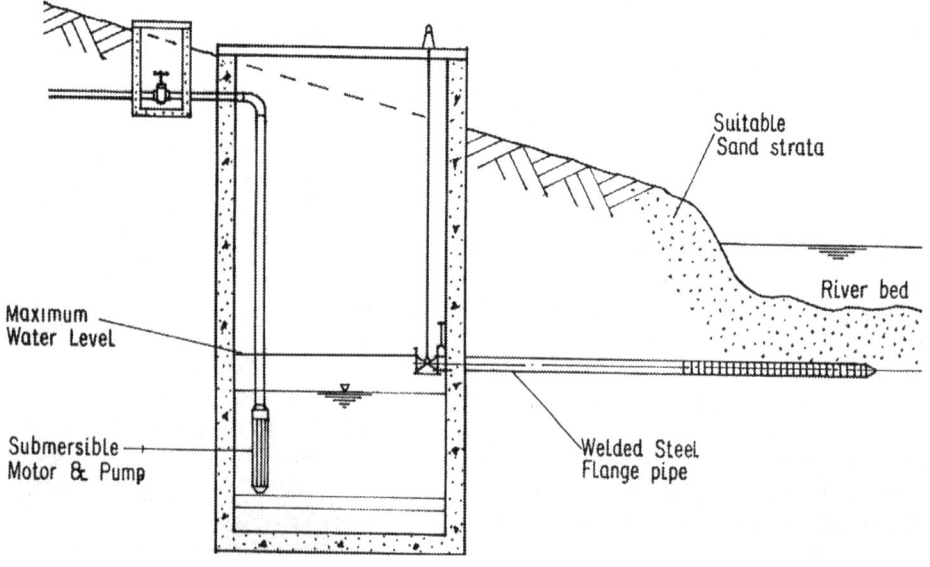

Figure 4.1D: A More Elaborate Intake Involving an Infiltration Gallery

4.4 Lake Intakes

4.4.1 INTRODUCTION

Deep and large lakes are subjected to temperature differences at different levels, a comfortable depth to draw from for domestic purposes would be 4–5 m below the surface. This would be too elaborate for the scale envisaged in this book. For agricultural purposes, the important thing is to locate the intake high enough above the lake bottom to avoid the entrance of silt; this applies to both deep and shallow lakes (see Figure 4.2A).

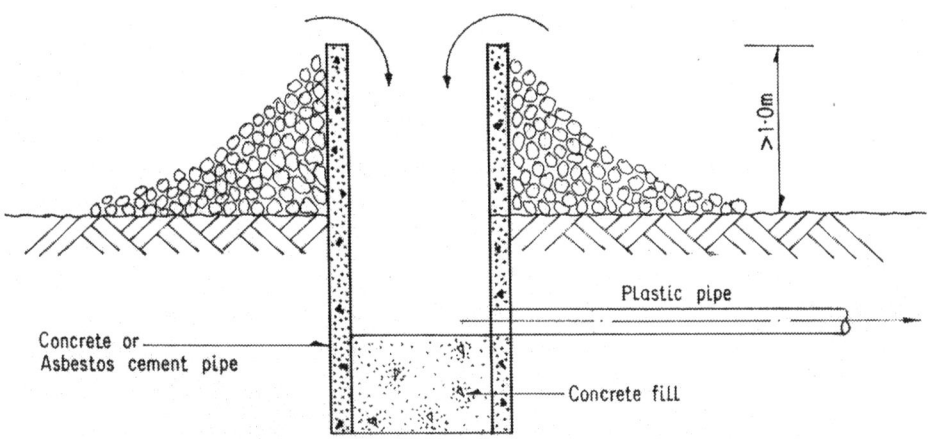

Figure 4.2A: Intake Structure at Bottom of Shallow Lake

4.4.2 TYPICAL INTAKE CONSTRUCTIONS

For a small community of about 1000 people (or fewer people with additional irrigation requirements), an intake requirement of about 1.4 l/sec. (84 l/min or 5,040 l/hr) would provide enough water for 4 times the requirement based on a per capita consumption of 30 l/day. For this purpose, a 150 mm (6") intake pipe would be sufficient to keep the intake velocity into the intake below 0.1 m/sec. For such small-capacity intakes, a simple flexible hose arrangement which draws the water into a well from where it is pumped out may be used (see Figures 4.2B and C).

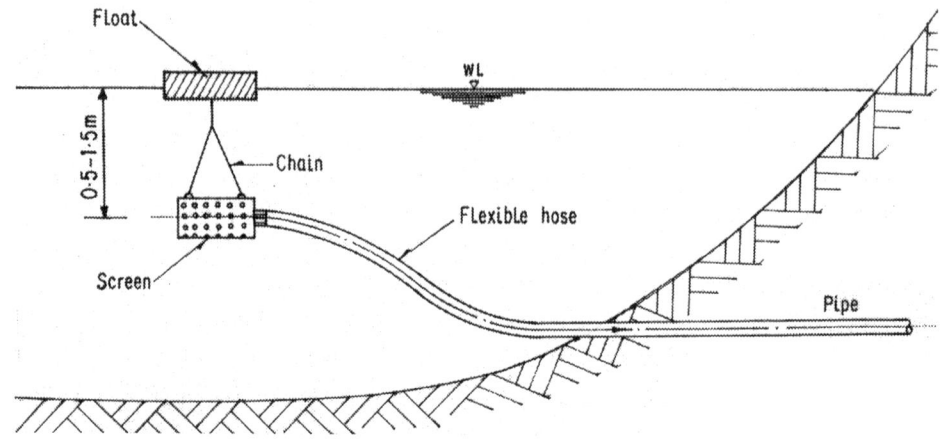

Figure 4.2B: Simple Water Intake Structure

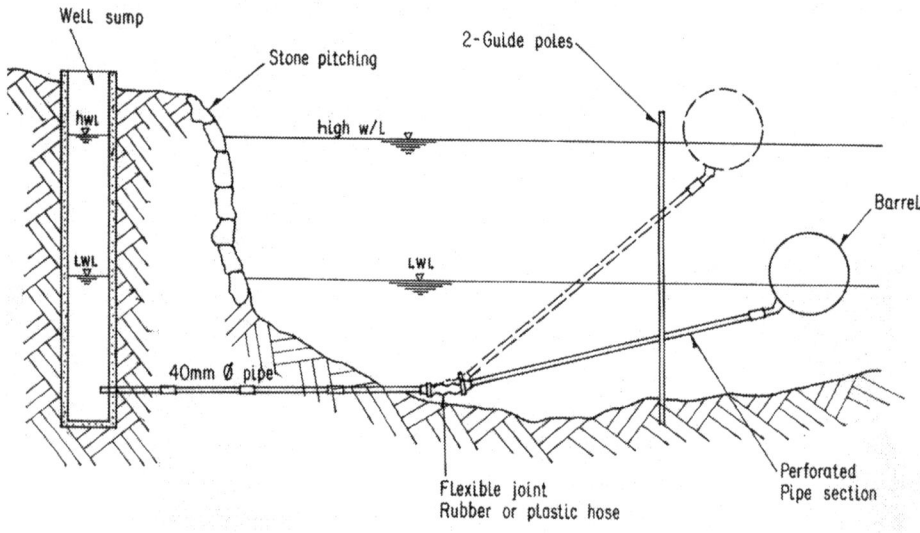

Figure 4.2: Float Intakes for Small Communities

4.5 Dams and Valley Tanks

4.5.1 INTRODUCTION

Apart from direct pumping or diversion from lakes and rivers which are more feasible in areas with perennial rivers and natural lakes, the major focus of surface water resources development is on dams and valley tanks (or dugouts).

4.5.2 PONDS

Ponds could be artificial and natural. Artificial ponds are formed by digging holes along or close to the flow channel, thereby trapping some water for use after the run-off or flood recedes. They could also be incidental depressions formed by borrowing for other purposes, e.g. sand, gravels and even quarrying for building and roadworks. Natural ponds are depressions, which eventually trap water as artificial depressions do.

Ponds are generally of limited capacities and usually within a farmer's holding or for village use; because of this limited capacity, ponds generally dry up rather early in the dry season.

Ponds are generally sources of poor quality water – physically and chemically – because of the turbid surface run-off which collects in them. They are also used for washing (laundry) and bathing, and people wade in the water to reach depths where their pots, buckets etc. for drawing water can easily collect enough and in search of less turbid water. Because of these inherent inadequacies and susceptibility to intense pollution, they are not encouraged as sources of domestic water as they are potential sources of epidemics – they harbour pathogens of bacteria, viruses and parasites, which account for innumerable diseases and even deaths especially among children. Primarily for this reason, they are not encouraged especially for domestic purposes.

In practice, it is not possible to prevent them from being polluted because they are generally situated at the lowest points of their surroundings; all the village drainage also finds its way into the pond sooner or later. However, if and where they are to be used for domestic purposes, the following steps should be taken:

- protection of the watershed from human and livestock activities – to prevent pollution
- minimise erosion – to reduce the amount of sediment and hence turbidity.

These can be achieved by fencing the watershed.

4.5.3   VALLEY TANKS AND DUGOUTS

These are more elaborate. Like ponds, they are excavated near the flow channel and filled by diverting the flow into them. Valley tanks usually have outlet structures and spillways while dugouts are only a little better than ponds but have no outlets and no spillways.

4.5.4   DAMS

At the extreme end of the surface water development scale are dams, which have been defined elsewhere. The procedures for developing dams and valley tanks are discussed from chapter V onwards.

Chapter V
REQUIRED STUDIES AND INVESTIGATIONS FOR DAMS AND VALLEY TANKS

5.1 Identification of the Need for Impounding Water

The need for developing water resources is generally as a result of demand exceeding the naturally available resources. Surface water is the water that occurs on the surface of the earth, generally from rainfall in tropical climates; it flows from several areas and finally concentrates in defined courses (channels, valleys, rivers etc.) until it collects in artificial reservoirs (dam reservoirs, valley tanks), lakes and, finally, in lagoons and seas. Surface water is more easily accessible to human beings, hence its development is usually the first option, especially for communal and large-scale exploitation, unless it is not readily available, as in arid and semi-arid climates.

The purpose of development in rural communities is usually for:

- agricultural purposes, i.e. crop production (irrigation) and livestock watering – fishery is always an added advantage,
- human domestic needs.

These purposes could be combined in which case one has a multi-purpose objective.

After defining the purpose(s) for which the development is envisaged, the procedure for further action is divided into three phases or stages: Pre-construction, Construction and Post-construction.

5.2 Studies, Design and Construction of Dams

Studies/activities which lead to the construction of dams (including valley tanks) include:

a. pre-design activities – mainly site selection, hydrological studies, field survey and soil investigations
b. design, including detailed design considerations and
c. calculations leading to the preparation of Bills of Quantities, Bidding Documents, Bid Analysis and Evaluation, Contractor Selection and Construction/Supervision.

The rest of this chapter discusses pre-design studies/activities, while chapters VI and VII deal with b. and c. (above) respectively.

5.3 Site selection

For the purpose of dam construction, a good site should satisfy all or at least most of the following criteria:

5.3.1. TECHNICAL CRITERIA

A.
  i. there must be enough water (run-off) to fill the intended reservoir.[3]
  ii. it must be possible to have the maximum possible storage with the smallest possible dam (embankment). In the case of rehabilitation, the steeper the side slopes of the valley above the existing dam, the greater the potential for increasing the reservoir capacity by raising the embankment (and the spillway correspondingly), but where desilting has to be done, the smaller the amount of excavation for maximum storage the better, i.e. relatively simple rehabilitation requirements. Embankment raising/spillway repairs could be a better option than desilting if the conditions make this feasible. (See section 6.3).
  iii. the site must afford a sound foundation and an impervious reservoir.
  iv. there must be suitable construction materials nearby to minimise costs.
  v. the site should be as close as possible to the site where the water is required to minimise transportation costs.
  vi. it must be possible to keep the impounded reservoir reasonably free from pollution, i.e. possibly upstream of the nearest human settlement; this is to encourage low catchment erosion hazards leading to rapid siltation – this is ultimately to reduce maintenance problems and so encourage sustainability over a long period of time.
  vii. it must be possible to have a good spillway site where the construction cost can be minimal.
  viii. the possibility of delivering water by gravity to irrigable areas, cattle watering points and delivery to humans downstream of the dam.
  ix. using the livestock per capita consumption (25–30 l/hc/day) versus available run-off, and a maximum walking distance of 5 km, it must be possible to calculate the livestock density (and human population) that can be watered from the intended water point for a projected period of time, up to year 2050 for example. For a multi-purpose scheme, irrigation water requirements could be added before the scheme requirement is arrived at.

---

[3] The water reaching the reservoir on a non-perennial stream depends on
  a. amount of rainfall;
  b. the size of the catchment; and
  c. Run-off – the proportion of rain water which becomes overland flow. Steep rocky slopes and a well-drained stream bed are preferable as they generate maximum run-off. Similarly, the presence of springs within the catchment and/or reservoir is of considerable advantage as it encourages smaller reservoirs for year-round supply.

B. Based on the foregoing, a good dam site to look for is one where a broad valley upstream suddenly narrows down. A dam built at such a location will be of minimum length and the broad valley upstream will provide the maximum storage. Similarly, the most economical site will usually be in a steep-sided valley (steep cross-section), where the water runs slowly, i.e. a flat longitudinal section. In summary, the faster the ground rises perpendicular to the stream, the shorter the dam would be and the lower the construction cost. Similarly, the gentler the slope along the stream bed, the further upstream will be the impounded water and the larger the reservoir for the same dam height. Most of these criteria are satisfied by the dam axis selected for Rwamuranda dam (figure 6.2).

C. For valley tanks (or dugouts), the major requirements are:
  i. availability of sufficient run-off to fill the intended reservoir i.e. (i) above and
  ii. the presence of an impervious stratum below the reservoir bottom. But when in dire need, clay lining (or blanket) could be placed at the bottom and on sides of the tank (reservoir) to improve the water-retention property of an otherwise unsuitable reservoir bottom stratum.

5.3.2. NON-TECHNICAL CRITERIA

a. These could be economic or political criteria which sometimes take precedence in the case of marginal sites.

b. Availability of encumbrance-free land on which the facilities are to be developed is a major factor in selecting a site that is technically suitable; an encumbrance-free but less technically suitable site could be selected in preference to a technically suitable site on which there are encumbrances.

5.3.3. FIELD SURVEY AND INVESTIGATIONS.

Once a site or sites have been identified, the selection made based on the criteria listed above is subject to confirmation after the necessary studies and field surveys have been carried out. The major studies and surveys are:
  i. hydrological studies
  ii. topographical survey at a suitable scale and
  iii. geo-technical investigations.

These surveys could be carried out simultaneously but it is customary to commence the first two simultaneously while the third one trails behind them. It should be noted that the topographical maps produced by the surveyor are very useful tools for all further work.

5.4 Hydrological Studies

5.4.1 RAINFALL/RUN-OFF RELATIONSHIP

Irrespective of the type of structure envisaged, hydrological study is a sine qua non; the overall purpose of hydrological studies is to determine the amount of water available at the proposed location (site). It is a study of the interrelationship between rainfall – the source of water – atmosphere (evaporation) and the catchment area (the land, vegetation, topography etc.) which determines how much of the rainfall eventually becomes available in form of run-off at the planned reservoir. It is from the outcome of this study that it is concluded whether there would be enough water to justify the development efforts and the reservoir capacity in relation to the water requirement.

For a given project site, the Hydrological Study is approached as follows:

5.4.2 DESCRIPTION OF THE PROJECT AREA
(in terms of rainfall and related weather and geographical characteristics that could affect water availability).

For example, the scheme lies within the dry belt part of Mbarara District, characterised by a long-term mean annual rainfall of about 990 mm. The rainfall records in the area are based on extrapolations from the rainfall records at Mbarara Meteorological Station, the only station with fairly long records (1966–1977 and 1980–1994) within the South-west region of Uganda. According to these records, the mean annual rainfall varies from about 438 mm to 1108 mm, with a mean annual value of 937 mm for the two periods mentioned above. In years like 1997–1998 when the El Niño weather phenomenon enveloped the whole of the East African region, the mean annual rainfall went up to 969 mm. The bimodal rainfall distribution of March to May and September to November (see tables 5.1A & B) has some limited groundwater recharge possibility i.e. rainfall in excess of evapotranspiration, only in March/April and October/November, with the latter period generating the bulk of the run-off. However, because of the apparent lack of continuity in the rainfall records and the total absence of run-off records since 1977, run-off and flood estimates based on these figures must be treated with extreme caution since there are no run-off and flood records to compare them with.

The area lies primarily in the rain shadow of the Kabula Hills and therefore falls within the low rainfall belt of the country, thereby having the characteristics of a forced semi-arid climate. It is underlain by the Basement Complex.

According to earlier studies (Gaf Consults Ltd.[4] and Fatokun, J[5]), the Lake Mburo Area falls entirely within a derived semi-arid region with average annual rainfall of 600–1000 mm; the project area therefore falls within the Lake Mburo belt.

There are no perennial rivers in the area, all sources of surface water are short-lived, discharging quickly into swamps, while the dotted pools they leave behind due to relief limitations dry up early in the dry season in response to the very high evapotranspiration.

TABLE 5.1A MEAN MONTHLY/ANNUAL PRECIPITATION AND EVAPO-TRANSPIRATION DATA FOR LAKE MBURO NATIONAL PARK AND ENVIRONS (1968–1977)

Para	Jan	Feb	Mar	Apr	May	Jun	Jul	Aug	Sept	Oct	Nov	Dec	Ann
PPT	57	70	114	130	65	28	34	69	101	125	137	63	993
ET	94	86	110	92	97	100	107	114	107	100	92	94	1193
PPT–ET	0	0	0	38	0	0	0	0	0	25	45	0	112

SOURCE: MBARARA METEOROLOGICAL STATION, KAKOBA

PARA = Parameter
PPT = Mean monthly/annual total precipitation in mm.
ET = Mean monthly/annual evapotranspiration in mm.
HNP = PPT–ET = Hydrological Net Precipitation
= precipitation in excess of evapotranspiration

The wind speed, estimated at 2.5 m/s at 0900 hrs and doubling at 1500 hrs with the high mean annual temperature of about 25 degrees Celsius, combine to create a very high annual evapotranspiration of about 1193 mm, ranging from 1100–1560 mm and giving an almost uniform monthly value of about 100 mm.

This description gives the reader a general idea of the inherent water-related potentials of the project area and more or less sets the stage for the following sections – assessment of the water resources of the area.

5.4.3 ASSESSMENT OF SURFACE WATER POTENTIALS

Based on the physical characteristics of the project area, its surface water potentials are assessed as follows:

i. Catchment Characteristics

The reliability of surface water storage at any point within a catchment area is a function of the rainfall/run-off relationship of the catchment, i.e. it depends on the amount of rainfall, the vegetative cover, nature of the soil and the slope on the catchment. These are generally referred to as the Catchment Characteristics.

---

[4]The Study of Water for Livestock and Domestic Use and the Related Socio-economic and Environment Issues in Lake Mburo Park and Environs, February 1993.
[5]SWRARP, The Water Component, Main report, July 1995 and Livestock Services Project, The Water Component, Main Report, March 1996.

ii. Hydrological Net Precipitation (HNP = PPT–ET)

The HNP given in Table 5.1A is an indication of the run-off potential of virtually all the catchments in the project area. The potential evaporation was taken as 0.8 of pan evaporation, (based on experiences from similar semi-arid climates – see footnote no. 5), giving a uniformly high annual Evaporation of 1320–1560 mm (3.6–4.3 mm/day). Run-off is possible only in April, October and November/December, with values ranging from 38 mm in April to 45 mm in November/December. There are variations in the length of the rainy season as well as in annual values within the area.

iii. Estimation of Run-off

The potential run-off of a catchment is a function of the rainfall it receives, the catchment characteristics and evaporation. There are a few methods of doing this, common among which are:

1. Oldest Inhabitant

    This is the crudest way of doing it, especially in very remote locations where there are no rainfall/run-off records. The engineer asks questions from the old people in the area regarding the level of water in the valleys, streams etc., the duration of the high water level and time of the year when it usually occurs, last occurred and its regularity, e.g. yearly. The answers to these questions from the oldest person or persons give a good indication of the catchment yield.

2. Run-off as a Factor of Rainfall

    Run-off is taken as a percentage of mean annual rainfall. This approach is not very good because of the wide variability of rainfall distribution within the year, but it is better than no assessment at all. The approach is used widely in Ghana, e.g. 10% is the figure generally used for the Upper East Region (UER). The FAO also considered this subject in detail in Ghana and decided to use 10% of mean annual precipitation after extensive fieldwork (Irrigated Agriculture in the Upper and Northern Region, Report No TA 2484, FAO, Rome, 1968).

3. US Bureau of Reclamation Approach

    The method uses 17% of mean annual rainfall as the annual run-off over the catchment area, based on Design of Small Dams (1st ed., 1960). This method is presumed to have taken most of the catchment characteristics into account and adopts this amount as reliable for design purposes to provide enough run-off to fill small dams.

4. Measurement of Stream Flows and Catchment Rainfall

    This is the most accurate method of obtaining run-off data, but the required data – rainfall and stream flow records – are generally not available, especially for small catchments. For large catchments (>5,000 km^2), run-off corresponds to about 7–10% of rainfall; 7.3 and

9.4% were obtained for catchments of 9,500 and 10,600 km² respectively in Ghana (personal experience).

After plotting measured run-off against corresponding rainfall figures from three different hydrologically homogeneous catchments in Western Nigeria, the following Regression Equations were found to fit Rainfall/Run-off relationships reasonably:

a. Assuming a linear relationship, $Q = 0.584R - 540$ mm[6]
   (this equation does not apply where $R < 925$ mm).
b. Assuming a curvilinear relationship, $Q = 0.85(R-1025)$ mm[7] or $0.85(R-41.0)$ ins.
   This equation applies only for $R > 1025$ mm or 45.0 in.

In these equations, Q = Run-off, R = Rainfall.

5. Ven Te Choy recommends 1,000 m³ of run-off for every 20 ha of catchment in semi-arid climates (Handbook of Applied Hydrology, 1964).

6. Working in a semi-arid area of Central and South-western Uganda, this author after extensive field work and literature review, adopted a catchment yield of 1,000 m³ of run-off per 40 ha of catchment. Most of the reservoirs designed on this basis were filled within one year of construction and remained full year in year out thereafter.

For comparison, the catchment yields obtained for four different catchments and five different methods are presented in Table 5.2.

This wide variability suggests that run-off estimation in the absence of measured run-off data is only an intelligent guess. Compared with measured values where data were available, some of these methods underestimate while others overestimate the potential run-off. But since experiences have been reported in many parts of semi-arid regions, it can be safely said that 10% is a good figure to adopt for planning purposes; it most closely approximates the stream gauging/rainfall records approach and provides enough run-off to satisfy the water requirements on most projects which were based on it.

---

[6] Scot Wilson Kirkpatrick & Partners, July 1977: Iseyin and Oke-Iho (Western Nigeria) Water Supply Scheme, Hydrological Studies

[7] Tahal Consultants (Nig.) Ltd., 1973: Ilorin Water Supply Extension – Planning Report for the Kwara State (Nigeria) Water Corporation.

TABLE 5.1B MBARARA RAINFALL (PPT MM) FOR THE PERIOD 1980–1994

YEAR	JAN	FEB	MAR	APR	MAY	JUNE	JULY	AUG	SEPT	OCT	NOV	DEC	ANN'L
1980	0	27.6	96.1	83	0	0	0	0	0	27.7	192.7	10.8	437.9
1981	43.2	48.2	72.5	192.7	84.4	23.9	28.9	142.5	112.2	134.8	95.3	44.3	1022.5
1982	23.8	51.5	52.7	82	122.2	39	17.7	37.7	112.6	91.1	168.9	87.4	886.6
1983	28.2	35.5	63.2	138.7	37.7	15.9	29.8	94.2	135.1	287.1	133.5	52.3	1051.2
1984	21.1	65.4	70.5	106.7	0	0	53	45.5	47.2	122.1	135.6	119.5	786.6
1985	46.6	4.7	50.7	89.2	26.4	3.7	6.1	1.3			71.3	0	
1986					62.4	5.4			95	152.2			
1987	83.8	34.7	100.5	133.3	87.2	38.1	15	8.3	118.9	51	106.3	54.6	831.7
1988	17.3	21.4	121	114.9	77.4	4.3	63.5	108	320.3	164.7	36.2	58.8	1107.8
1989	10.4	4.3	99.1	135.4	102.2	6	31	37.2	151.7	136.8	68	103.4	885.5
1990	33.6	92.1	139.4	131.1	23.4	0.9	0	68.9	100.4	117.9	146.3	107.4	961.4
1991	70.7	37.1	46.5	144.4	48.3	57.7	42.7	63.5	40.7	214.5	90.4	126.6	983.1
1992	13.8	37.4	116.1	121.2	36.4	42.3	29.4	10.8	77	43.1	95	48.6	671.1
1993	41.1	49.4	60.5	170.1	97.9	20.8	0	125.6	67.9	114.1	50.2	89.7	887.3
1994	100.2	40	108.9	86.4	120.9	6.1	6	45.7	34.8	138.7	156.1	104.4	948.2
MEAN[8]	37.5	41.9	88.2	126.2	64.4	19.6	24.4	60.6	101.5	126.4 13.5	113.4	77.5	881.6

SOURCE: MBARARA METEOROLOGICAL STATION, MBARARA

---

[8] 1985 & 1986 excluded.

5.5 Determination of Scheme/Project Water Requirements and Storage Capacity

The objective of a water resources development project is to make water available on a sustainable basis. Most rural-based water resources schemes are multi-purpose, making water available for human and livestock needs as well as crop production – irrigation.

A community in the savannah area of Oyo State, Nigeria has the following human and livestock populations: 10 households of an average membership of 5, each having 20 head of cattle. There are 90 sheep and goats altogether. Other characteristics of the community are as follows:

TABLE 5.2   CATCHMENT YIELDS BY DIFFERENT METHODS

LOCATION	MEAN ANNUAL RAINFALL (MM)	CATCHMENT AREA (HA)	CATCHMENT YIELD BY DIFFERENT APPROACHES MILL. M³ (MCM)				
			10% OF MEAN ANN. RAIN.	17%, I.E. AFTER USBR	VEN TE CHOY – 1000 M³ PER 20 HA	J FATOKUN – 1000 M³ PER 40 HA	STREAM GAUGING-RAINFALL RECORDS
1. Sissili, UER, Ghana	1044	950,000	991.8	1,679.6	47.5	95.0	695.7
2. Kulpawn, UER, Ghana	1044	1,060,000	1,106.7	1,881.3	53.0	106.0	1,000.3
3. Kanyaryeru, West Uganda	937	2,337	2.2	3.7	0.029 (29,213)	0.058 (58,425)	data not available
4. Akayanja, West Uganda	937	5,030	4.7	8.0	0.63 (62,875)	0.126 (125,750)	data not available

There is a very large primary school with an enrolment of 1150 pupils and a secondary school of 402 pupils; these schools are owned and run by missionaries for a very large catchment area. The rainfall in the area is both erratic and unreliable and the distribution is uneven, hence, crop production is limited to the rainy season of about five to six months – April to October, recording a mean annual figure of 930 mm. There is an urban centre within some 50 km that offers a good market for an all-year round vegetable production. The community therefore wishes to take advantage of a proposed dam-based water scheme to produce vegetables and some household food during the dry season. It is presumed that there is good land downstream of the proposed dam site for dry season cropping, if water can be provided. Would the aspirations of the community be realised if the government was approached for assistance?

Solution
Length of dry season: November to March/April – 165 days.
a. Water Requirements to meet:
   i. Human and Livestock Needs as tabulated in Table 5.3:

TABLE 5.3   HUMAN AND LIVESTOCK WATER REQUIREMENTS

POPULATION	DAILY WATER REQT. (LIT PER CAPITA PER DAY – LCD)	TOTAL SEASONAL REQUIREMENT (M^3)
1. Human 50 2. Schools 1,552 3. Cattle 200 4. Sheep/goats 90	25 20 30 7.5	194 3,571 930 105
	Sub-Total Allowance for unforeseen increases (30%)	4,810 1,443
	TOTAL	6,253

ii Crop Water Requirements
It is suggested that each holding be 1 ha, and in line with the proposal to grow vegetables for the nearby market and some food crop for household consumption, a cropping pattern of 80% vegetables and 20% maize (0.8 ha for vegetables and 0.2 ha for maize) will meet that aspiration.

NOTES: Irrigation water requirement depends on crop water requirement, which in itself depends on evaporation, crop type, amount of vegetative cover on the field etc. In line with this, crop requirement for a certain period is obtained from the equation

$$E_{tcrop} = E_o \times k_o$$

where E_{tcrop} = crop requirement
E_o = open water evaporation
k_o = crop factor (coefficient) which varies with the stage of crop on the field.

Tables 5.4 and 5.5 show the derivation of the crop water requirements of maize and vegetables, based on the adopted cropping pattern of 0.8 ha and 0.2 ha respectively.

From these tables, the season's requirement for vegetables and maize per holding of 1 ha is approximately equal to 11,000 m³ as derived below:

CROP	IRRIGATION APPLICATION DEPTH (m)	WATER REQUIREMENT (m³)
Vegetables (0.8 ha)	1.067	8,536 (0.8 ha x 1.067)
Maize (0.2 ha)	1,154	2,308 (0.2 ha x 1.154)
Total Requirement		10, 844

(i.e. approximately, 11m³/ha)

Thus the total irrigation requirement for the 30 ha scheme = 11,000 m³/ha x 30 ha = 330,000 m³.

iii.

TOTAL SCHEME WATER REQUIREMENT (i) + (ii)	(i) 6,253	(ii) 330,000 m³
		= 336,253 m³
	Add 20% for seepage losses from the reservoir	67,250
	TOTAL	403,503 m³

b. Would the catchment yield support (provide) this water requirement?

The hydrological study of the catchment showed that the catchment can yield 3.2 million cubic metre (m cm) of water in an average year (assessment similar to the figures shown in Table 5.2). The catchment is therefore able to support the scheme.

c. The scheme requires less than 15% (actual is about 13%) of catchment yield. Is it expedient to build a dam only for this storage requirement?

The answer to this question borders on site conditions and ability to pay the cost.

i. Site Conditions. It is always expedient to develop the full potentials of a dam site at once because it generally becomes very expensive to increase storage capacities at future dates. From this point of view, it is better to build a dam with maximum storage capacity once and for all, and it is possible in this case. Because the dam axis is a very good one (see Site Selection Criteria), the maximum possible storage capacity was adopted, 2.6 m cm (See Figure 5.1).

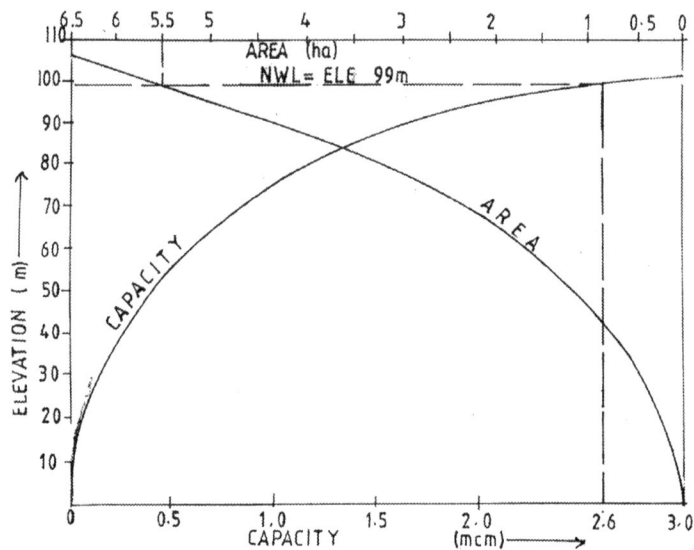

Figure 5.1: Area/Capacity Curves of Sepeteri Dam

ii. Can the funding agency provide the necessary funds? If yes, the parameters of the bigger dam/reservoir (2.6 m cm) option are built into the final design and bidding document, otherwise, the smaller dam/reservoir (403,503 m^3 option is adopted.

As an example of the approach to solving this kind of problem, a case study is presented here: Wabale Dam in Nakasongola District, Uganda.

After the dam construction had already started, following the design which was based on the Engineer's initial brief, the user community got to know the details of the benefits that would accrue to them, which apparently left out urban water supply. The community wanted to take advantage of the dam, which was located very close to the town and consequently requested that the engineer explore the possibility of increasing the reservoir capacity, the additional costs of which they were prepared to pay. What follows is the actions taken by the engineer. The project manager to whom the engineer addressed his report brushed aside the report, claiming that he didn't want anything to tamper with the completion schedule of the project. The report nevertheless, gives an idea of how to go about this kind of problem in water resources development.

Ref: LSP/WC/JF/075   14 April 1998

The Director of Animal Resources

Dear Sir

LIVESTOCK SERVICES PROJECT – THE WATER COMPONENT: CONSTRUCTION/REHABILITATION OF DAMS AND VALLEY TANKS

THE ADDITIONAL COST OF INCREASING THE RESERVOIR CAPACITY OF WABALE DAM, NAKASONGOLA DISTRICT

During the orientation seminar held on 18 February 1998, the officials of Nakasongola District who attended requested for the additional cost of increasing the Reservoir Capacity of Wabale Dam with a view to paying that cost so that the dam could supply water for domestic purposes to the nearby Nakasongola town.

Determining this cost required additional study and computations which have now been completed and the parameters of the new dam are as follows with the corresponding figures for the present dam in brackets:

Height	=	8 m (6.25 m)
Embankment length	=	460 m (402 m)
Total reservoir capacity	=	340,000 m^3 (185,000 m^3)
Additional capacity	=	155,000 m^3

The additional cost, based on the present contractor's rates is Sh. 76,917,830 (Table 5.6).

If the project is willing to accommodate this proposal, it will be necessary to convey this to the district with dispatch and take any further necessary action to enable us instruct the contractor accordingly. It will be cheaper to carry out the necessary foundation works before the contractor goes far from the present ground level.

Yours faithfully,

Sgd.
WATER RESOURCES DEV. CONSULTANT

TABLE 5.4 WATER REQUIREMENTS OF MAIZE (MM PER HALF MONTH PERIOD)

MONTH	PERIOD	E_T	k_o	$E_{T CROP}$	LAND PREPARATION	PERCOLATION	GROSS	$R_{EFF.}$	NET	FARM TURNOUT	CANAL HEAD
January	1	70	1.05	74	0	0	74	15	59	79	113
	2	76	1.04	79	0	0	79	1	78	104	148
February	3	82	1.05	86	0	0	86	2	84	112	160
	4	78	1.05	82	0	0	82	2	80	106	152
March	5	80	1.06	85	0	0	85	8	77	103	147
	6	91	1.08	98	0	0	98	8	90	120	172
April	7	91	0.37	34	0	0	0	0	0	0	0
	8	87	0.00	0	0	0	0	0	0	0	0
May	9	84	0.00	0	0	0	0	0	0	0	0
	10	83	0.00	0	0	0	0	0	0	0	0
June	11	77	0.00	0	0	0	0	0	0	0	0
	12	70	0.00	0	0	0	0	0	0	0	0
July	13	66	0.00	0	0	0	0	0	0	0	0
	14	64	0.00	0	0	0	0	0	0	0	0
August	15	62	0.00	0	0	0	0	0	0	0	0
	16	61	0.00	0	0	0	0	0	0	0	0
September	17	62	0.00	0	0	0	0	0	0	0	0
	18	63	0.00	0	0	0	0	0	0	0	0
October	19	67	0.00	0	0	0	0	0	0	0	0
	20	71	0.00	0	50	0	0	0	0	0	0
November	21	70	0.00	0	0	0	50	3	47	63	90
	22	66	0.40	26	0	0	26	3	24	32	46
December	23	65	0.41	27	0	0	27	1	26	35	50
	24	68	0.60	41	0	0	41	1	40	54	77
		1,753		616	50	0	632	27	606	808	1,154

NOTES:
1. Field Irrigation Efficiency = 75%.
2. Canal Conveyance Efficiency = 70%.
3. The crop is sown or planted 0 days after beginning of period 22, and is harvested in period 7.
4. Irrigation is stopped 10 days before harvest.
5. Values of effective rainfall (R_{eff} – amount of rainfall retained in root zone which irrigation needs not supply) are calculated using USDA method. E_t = open water evaporation; k_o = crop factor (coefficient); E_{tcrop} = crop evapotranspiration.

TABLE 5.5 WATER REQUIREMENTS OF VEGETABLES (MM PER HALF MONTH PERIOD)

MONTH	PERIOD	E_T	K_o	E_{TCROP}	LAND PREPARATION	PERCOLATION	GROSS	R_{EFF}	NET	FARM TURNOUT	CANAL HEAD
January	1	70	1.05	70	0	0	70	0	70	93	133
	2	76	1.05	79	0	0	79	0	79	105	150
February	3	82	1.05	86	0	0	86	2	84	112	160
	4	78	1.06	82	0	0	82	2	80	107	153
March	5	80	1.09	87	0	0	87	6	81	108	154
	6	91	0.00	0	0	0	0	0	0	0	0
April	7	91	0.00	0	0	0	0	0	0	0	0
	8	87	0.00	0	0	0	0	0	0	0	0
May	9	84	0.00	0	0	0	0	0	0	0	0
	10	83	0.00	0	0	0	0	0	0	0	0
June	11	77	0.00	0	0	0	0	0	0	0	0
	12	70	0.00	0	0	0	0	0	0	0	0
July	13	66	0.00	0	0	0	0	0	0	0	0
	14	64	0.00	0	0	0	0	0	0	0	0
August	15	62	0.00	0	0	0	0	0	0	0	0
	16	61	0.00	0	0	0	0	0	0	0	0
September	17	62	0.00	0	0	0	0	0	0	0	0
	18	63	0.00	0	0	0	0	0	0	0	0
October	19	67	0.00	0	0	0	0	0	0	0	0
	20	71	0.00	0	50	0	50	7	43	58	82
November	21	70	0.35	24	0	0	24	2	22	29	41
	22	66	0.35	23	0	0	23	2	21	28	40
December	23	65	0.48	33	0	0	33	1	31	41	58
	24	68	0.74	50	0	0	50	0	50	66	95
		1,753		534	50	0	584	23	560	747	1,067

NOTES:
1. Field Irrigation Efficiency = 75%.
2. Canal Conveyance Efficiency = 70%.
3. The crop is sown or planted 0 days after beginning of period 21, and is harvested in period 5.
4. Irrigation is stopped 5 days before harvest.
5. Values of effective rainfall (R_{eff} – amount of rainfall retained in root zone which irrigation needs not supply) are calculated using USDA method. E_t = open water evaporation; k_o = crop factor (coefficient); E_{TCROP} = crop evapotranspiration.

TABLE 5.6 WABALE DAM, NAKASONGOLA DISTRICT

ADDITIONAL COST OF INCREASING THE RESERVOIR CAPACITY FROM 185,000 M³ TO 340,000 M³

The additional cost to be incurred is mainly on earthworks and fencing, all other aspects remain as designed for the 185,000 m³ capacity. These costs are shown in the table below.

BILL	DESCRIPTION	UNIT	DISTR. PROPOSED	LSP (PRESENT)	DIFFERENCE	RATE SH '000	AM'T SH '000
1.01	Clear bush and grub top soil over dam and reservoir area	ha	10.5	10.5	–	–	–
1.02	Excavate foundation trench for dam core	m³	2,070	1,809	261	5.5	1435.5
1.03	Breach embankment	–	–	–	–	–	–
1.04	Borrow, cart, fill & compact good clay for dam core	m³	7,570	6,773	797	5.5	4383.5
1.05	Ditto for dam wall, using good murram	m³	34,060	20,649	13,031	5.5	71670.5
1.06	Cut spillway and cart to spoil/use	m³	7,958	9,725	–1767	5.5	–9718.5
1.07	Trim to proper design slopes	m²	10,396	7,300	3,096	0.35	1083.6
1.08	Provide, lay & compact lightly a 150 mm lateritic/gravelly material capping	m³	277	241	36	5.5	198
1.09	Top-soiling & grassing – provide and lay 50 mm deep top soil and grass as protective cover against erosion:	m²					
	a. Dam upstream slope		5,198	3,650	1,548	}	
	b. Dam downstream slope		5,198	3,650	1,548	} 0.5	471
	c. Spillway Control Section		300	300	–	}	
	d. 30m wide filter strip U/S of reservoir		7,246	9,300	–2,054	}	
	TOTAL BILL NO 1						69523.6
4.01	Fencing	m	1,633	1,530	103	3.9	401.7
	TOTAL ADDITIONAL COST						69,925.3

Add 10% Contingency 6,992.53
GRAND TOTAL = SH. 76,917,830

Considering the Kanyaryeru Resettlement Scheme, whose project area was described as a sample (5.4.2 above), the following was concluded:

Based on the hydrological studies of the project area, a catchment yield of 1,000 m^3 of annual run-off per 40 ha of catchment area was adopted, being twice the catchment area recommended for similar semi-arid areas,[9] to ensure that the run-off obtained would be reliable. It was also recognised that the run-off from relatively small catchments could be higher as seen for other catchments for which rainfall-run-off data are available – Table 5.2. For six locations in the Kanyaryeru Resettlement Scheme, the available volumes of water within their catchments as estimated above, were compared with the required Reservoir Storage Capacities and a decision was taken regarding which sites could be developed, depending on whether there was sufficient water within the catchment to fill the desired reservoir. The results are presented in Table 5.7.

---

[9] Ven Te Chow, (ed.), *Handbook of Applied Hydrology*, McGraw-Hill Inc., USA,1964.

TABLE 5.7 ESTIMATED ANNUAL CATCHMENT YIELD VERSUS REQUIRED STORAGE – KANYARYERU RESETTLEMENT SCHEME

SITE NAME	CATCHMENT AREA (HA)	REQD. STORAGE (M^3)	CATCHMENT YIELD (M^3)	REMARKS	RESERVOIR CONSTRUCTED (M^3)
1. Rwebikondo	243	13,057.2	6,073	Insufficient	6,805
2. Kanyaryeru (MO)	2,337	3,906	58,425	Sufficient	1. 5,805 2. 11,369
3. Rwengiri	639	9,873.5	15,578	Sufficient	11,935
4. Rwendama	620.4	7,300.5	15,510	Sufficient	11,935
5. Akayanja	5,030	6,990.5	125,750	Sufficient even for a dam	1. 5,805 2. 11,935
6. Mpangamu-shanju	490	5,160	12,245	Sufficient	11,369

In the absence of good dam sites in Kanyaryeru, all the reservoirs were to be valley tanks. For the sites under consideration, the objective was to construct valley tanks (reservoir) of storage capacities of about 5,000 m³ each. As seen in Table 5.7, one of the sites – Rwebikondo – did not have enough catchment yield to fill the contemplated reservoir capacity (volume). Under normal conditions, the site would have been dropped from further consideration, but because the resettlement site it was intended to serve was already in place, the valley tank was constructed with the hope that in future, water would be 'imported' from a nearby catchment which had enough to spare, i.e. the Kanyaryeru (MOW) site. (Table 5.7).

The depth of the tanks is then fixed, considering the effect of evaporation. In this case, with a mean annual evaporation of some 1,193 mm compared with an annual precipitation of about 937 mm, it is clear that the single largest water loss from a storage reservoir in this area is through evapotranspiration. Because evaporation is less for deep reservoirs, an average reservoir depth of 3.5 m (range: 3–4.2) was adopted for the design of the valley tanks in question.

In the case of dams, the reservoir capacity is a function of dam height and the surface area of land inundated. Depending on how much area the project owner (farmer or community) is prepared to lose (i.e. put under permanent inundation), the reservoir capacity could be varied by increasing or decreasing the height of the normal water level (NWL) or full supply level (FSL) and consequently the dam height. This is usually done with the help of area/capacity curves.

5.6 Determination of Water Levels and Dam Height

Calculation of Reservoir Areas.

From the contour lines on the topographical map of the proposed reservoir, the areas of successive plains are calculated and plotted against the capacities obtained for each section (slice) of the intended body of water. The areas could be calculated manually, using triangulation method or with the use of a Planimeter. This author used the two methods and found that the results were very similar – (see later in this section).

The volume (or capacity or storage) for successive 'slices' of a reservoir (a trapezoidal 3-dimensional object) is computed from the equation:

i. Capacity (Storage) $S_j+1 = [S_j+(h_j+1-h_j)(A_j+A_j+1)]/2$ ...1
   where S_j+1 = storage at level 1(one) contour interval above the previous level
   S_j = storage at elevation j
   h_j = height above reservoir bed at elevation j
   h_j+1 = height at 1 (one) level above level j

ii. Using the CONIC method for calculating the areas of conical objects
   $(A_j+A_j+1)/2$ in equation 1 above is replaced with: $[A_j+A_j+1+/A_j x A_j+1]/3$,
   thus $S_j+1 = S_j+(h_j+1-h_j)[A_j+A_j+1+/A_j x A_j+1]/3$ ...2

Equations 1 and 2 give slightly different values as shown in Table 5.8 (columns 4 & 5). For planning purposes, it is advisable to use equation 1 because it consistently gives higher values

TABLE 5.8: AREA/CAPACITY DATA FOR RWAMURANDA, MBARARA DISTRICT, UGANDA

LEVEL	ELEVATION (M)	AREA (M²)	CAPACITY (STORAGE) (M³)	
			EQUATION 1	EQUATION 2
1.	26.0	-	-	-
2.	26.5	2,500	625	416
3.	27.0	6,480	6,870	2,583
4.	27.5	10,656	7,154	6,824
5.	28.0	19,630	14,726	14,282
6.	28.5	32,312	27,711	27,137
7.	29.0	44,537	46,923	46,628
8.	29.5	55,572	71,951	71,244
9.	30.0	69,914	103,322	102,417
10.	30.5	76,205	139,852	138,936

The area/capacity curves for this example are shown in Figure 5.2.

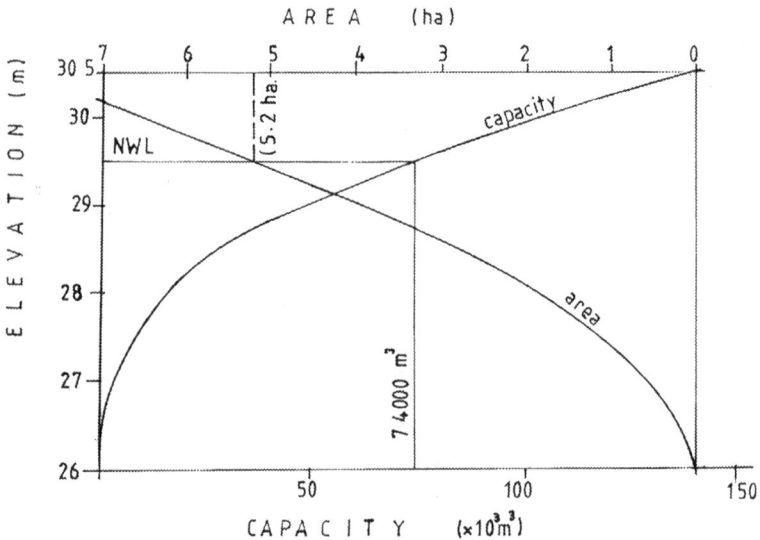

Figure 5.2: Rwamuranda Dam – Area/Capacity Curves

From the hydrological study of the catchment of this site, it was concluded that the catchment yield was between 12,600 m³ and 63,100 m³, based on different approaches; an average catchment yield of 35,800 m³ was adopted. Table 5.9 shows this range along with the figures obtained for other sites in the same project.

The reservoir's FSL or NWL is then chosen to obtain the storage capacity arrived at on the basis of comparing water requirement with catchment yield; in the case of Migeera Dam (Table 5.9 & Figure 5.3), the water requirement was 125,000 m³ and the catchment yield 161,900 m³. But the reservoir capacity was 185,000 m³, adopted to fully exploit the potentials of the site and to ensure that any inadequacies in the estimation of catchment yield did not inadvertently limit the reservoir capacity. This approach ensures that any high yields obtained in years of above-average rainfall could be progressively accumulated for use in years of severe drought. This 'total potential approach' would also ensure that the water requirements are met, and that the additional storage built into the reservoir would not be acquired at much higher costs when it would be mandatory to raise the capacities of water-retaining hydraulic structures at later dates when demands exceed the present capacities.

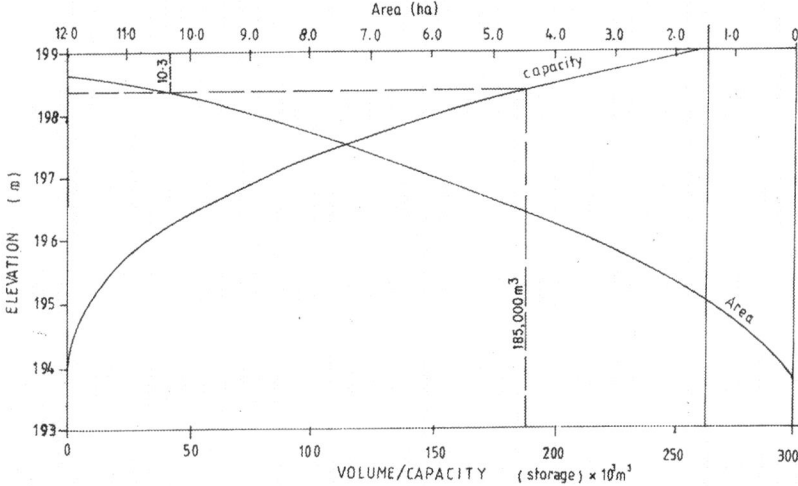

Figure 5.3: Migeera Dam Axis – Area/Capacity curves

In the detailed example in question (i.e. Rwamuranda Table 5.9 and Figure 5.2), the NWL was fixed at elevation 29.5 m (Figure 6.2), thus determining a reservoir capacity of 74,000 m³, i.e. twice the volume arrived at from catchment yield analysis.

The maximum water level (MWL) is generally determined by routing a flood of known magnitude through a full reservoir (filled to NWL). What is likely to happen when such a flood flows into the full reservoir? In the event that the dam is unable to contain the flood, what happens if it breaks? For the small dams considered here, there is little danger to human life if they break, but any crops and livestock that cannot escape are likely to be seriously endangered. Using the data below, let us see what could happen.

Length of dam = 500 m
Reservoir area at NWL (FSL) = 10 ha
Average ground slope in catchment = 1:100 (1%)

Free board of dam = 1 m
Maximum rain falling in 15 hrs. = 115 mm
Catchment area = 288 ha
Run-off = 40 mm (17% after USBR)
Outlet pipe = 225 mm diameter pipe, 45 m long

(i) Volume of water assumed to arrive at the reservoir within the rainfall duration (15 hrs)
  Run-off Volume = 288 ha x 10,000 m²/ha x 40 mm (40 mm = 0.04 m)
  = 115,200 m³

(ii) Volume stored above NWL –
  a. Reservoir Area = 10 ha
     Allowable safe rise in reservoir level = 50% of free board = 0.5 m
     Therefore volume = 0.5 m x 10 ha x 10,000 m²/ha
     = 50,000 m³
  b. Flood Volume stored in
     catchment upstream of reservoir = Area x length of extended perimeter
     (assume an area of 12.5 m² and extended perimeter of 10,548 m)
     = 13,185 m³
  c. Total Stored Volume = 50,000 + 13,185 m³
     = 63,185 m³

(iii) Hydrograph[10]
It is assumed that run-off begins to reach the reservoir 3 hours after rainfall starts and that peak discharge into the reservoir occurs 15 hours after rainfall commences and that the hydrograph has a straight line.

  Area of hydrograph = Volume of run-off
  Q/t = 276,480 m³
  Q = (2 x 115,200) / (15 x 360)
  = 4.27 m³/sec.

(iv) Reservoir Operation
  a. During the flood period (i.e. hour 3 to 18), the outlet pipe is open and discharging;
     Volume discharged = 270 l/sec (i.e. under the given head and diameter)
     = 97.2 m³/hr or 0.27 m³/sec
     This amount should be deducted from the base of the hydrograph as it does not remain in the reservoir, the revised hydrograph therefore has an area of 4.27 – 0.27 = 4.0 m³/sec.
  b. Spillway
     Using the broad crested formula (the spillway is actually a broad-crested weir):

---

[10] For details of Hydrograph Analysis and Flood Routing, see Wilson, E M and USB under References.

$$Q = CBH^2 \text{ m}^3/\text{sec}$$
Where $C = 1.65$ (weir constant)
H = height m
B = crest width (spillway width m)

Assuming a crest width of 10 m (i.e. spillway width), the following values of Q were obtained:

H (m)	Q (m³/sec)
0.1	0.5
0.2	1.49
0.3	2.64
0.4	4.13
0.5	5.78

This table shows that a 10 m wide spillway will accommodate the flood safely; it is unlikely that the flood would rise by more than 0.4 m above the NWL. The surcharge of 0.4 m is only required to give the spillway the necessary head (or drive) to spill. Surcharge is the maximum allowable rise in water level above the spillway control section.

In such cases, and for small dams, allowing an ample freeboard of at least 1 m would be sufficient to safeguard the embankment against overtopping in the event of unprecedented floods.

TABLE 5.9   ESTIMATED CATCHMENT YIELDS FOR SOME SELECTED SITES

DISTRICT/SITE	WATER REQ. (M^3)	CAT. AREA (HA) ($KM^2 = HA/100$)	CAT. YIELD (GAF '93 – I.E. 100 HA/1000 M^3) ('000 M^3)	CAT. YIELD (CHOW '64 I.E. 20 HA/1000 M^3) ('000 M^3)	CAT. YIELD (LSP – I.E. 40 HA PER 1000 M^3) ('000 M^3)	CAT. YIELD (AVERAGE) ('000 M^3)
MBARARA						
1. NYAKAHITA	152,000	-	-	-	-	-
2. RWAMURANDA	43,000	1,263	12,630	63,150	31,575	35,785
3. KYERA	40,000	2,484	41,340	124,200	62,000	75,847
4. KENWA	128,000	-	-	-	-	-
5. KISHINGURA	110,000	-	-	-	-	-
6. RWENSINDIZI	105,000	960	16,000	48,000	24,000	29,333
NTUNGAMO						
7. KIGAAGA	250,000	3,010	50,176	150,500	75,250	91,972
NAKASONGOLA						
8. MIGEERA	125,000	5,300	88,333	265,000	132,500	161,944
9. WABALE	30,000	1,950	32,500	97,500	48,750	59,583
10. BURULI–KAZWAMA (MAMBA SWAMP)	70,000	3,670	61,167	183,500	91,750	112,139
SEMBABULE						
11. KYAMBIDDE	120,000	500 ?	8,333	25,000	12,500	15,278
12. RWAMAKARA	135,000	2,600	43,333	130,000	65,000	79,444
MUBENDE						
13. DYANGOMA	408,000	1,000	16,667	50,000	25,000	30,000
14. LWEMITONGOLE	125,000	-	-	-	-	-
15. KASENSERO	26,500	-	-	-	-	-
KIBOGA						
16. NAKAKABALA	45,000	1,245	20,749	62,250	31,125	38,041
17. KASEIJERE (RANCH 7)	28,000	1,470	24,167	72,500	36,250	44,307
18. WABIKUNYU	12,000	1,100	18,333	55,000	27,500	33,611

5.7 Terms of Reference for Topographical Survey

Preamble

The purpose of this survey is to prepare the topographical maps of sites presumed as suitable dam/reservoir sites, based on visual reconnaissance inspection, and ultimately to use them to confirm the selection of such sites. The map will subsequently be used by the Design Engineer for his design of the dam wall plan, the spillway plan and the NWL. Other features of the dam and its appurtenant works may also be put on the map. The Surveyor should therefore take care to ensure that the Topographical map he produces represents the field levels accurately.

Specific Terms of Reference (TOR)
1. This survey is to be at scale 1:1,000 with contour intervals at 0.5 m.
2. The benchmarks used (at least two), should be located where they can be used for all levelling work during construction and should be clearly indicated on the topographical drawing (not within the proposed reservoir).
3. The parcel of land to be surveyed for dam axis and reservoir sites should cover from about 200 m downstream of the axis to about 1 km upstream, giving a total reservoir water spread area of about 10 hectares. The area under each contour line should be shown in hectares on the topographic maps along each contour line.
4. The presence of any special features and structures should be noted and clearly annotated; spot heights of important landmarks should also be indicated and annotated.
5. Where the dam axis has been marked on the ground and/or described in writing, the axis should be clearly pegged on the ground or otherwise marked for ease of identification by the geotechnical survey team. It should also be clearly shown on the drawing by means of a broken line.
6. The longitudinal profile of the dam axis as described in (5) above should extend far enough on both banks to show any saddle or sudden change in elevation that may serve as good location for the spillway.
7. For existing dam sites, the features of the embankment should be highlighted, including shape and elevations, and the present water level clearly defined by a closed contour line. The spillway should similarly be clearly highlighted down to the point of discharge into the valley downstream of the embankment.
8. The area of the entire catchment should be determined with the aid of the existing national topo sheets of scale 1:50,000 (or higher), usually obtainable from the National Survey Department.
9. The surveyor will prepare the original topo drawings on Ammonia paper

(size to be agreed with the engineer); he will deliver this original drawing together with two clearly legible printed copies for each site.

The above-listed specific requirements of this survey notwithstanding, the survey should be carried out according to the standard practice of the surveying profession.

Figure 6.1A and 6.1B and 6.2 are examples of topographical survey reports (map or drawing), based on this TOR, on which the designs of the respective structures were superimposed.

5.8   Soil and Geotechnical Investigations

i. Introduction

The purpose of this survey is to assess the suitability of the foundation of the selected dam axis to hold a water-retaining structure, and to determine the kind of foundation treatment required), the suitability of the soils around the reservoir and its environs as good materials for dam construction and water tightness of the reservoir floor. The survey is intended to identify and specify the best materials (murram) for dam construction in that vicinity.

To achieve the above-stated objectives, the standard practice is to carry out field and laboratory investigations and analysis/tests, the summaries of which are given below:

ii. Field Investigations

These are to be carried out in accordance with the British Standard Institution (BSI) 5930: London, 1981: 'Code of Practice for Site Investigation' or equivalent. Drilling along the dam axis and reservoir areas is to be as deep as 6.5 m and of 200 mm diameter. Standard penetration tests (SPT) and sampling of undisturbed and disturbed samples are to be carried out. The soil stratifications are to be shown and described up to 6.5 m depth or minimum 3 m, depending on the layers encountered. The location and depth of each borehole/test pit should be shown on the topo survey drawing prepared by others.

Termite hills around the dam and reservoir areas are good indicators of the locations of clay and good murram. Samples are to be collected from these locations and tested for suitability. Appropriate reports with charts, tables etc are to be written.

iii. Laboratory Analysis and Testing

This is to be conducted in accordance with BSI 1377: London, 1975: 'Method of Test for Soils for Civil Engineering Purposes'.

These tests include: classification i.e. sieve analysis, liquid limit, plastic limit, plasticity index (p.i.) and oedometer consolidation, compaction, permeability, triaxial compression and linear shrinkage.

iv. Results and Recommendations:

The field investigations and laboratory analysis should be interpreted and used in making specific recommendations on the foundation conditions of the dam axis, the water retention property of the reservoir and the suitability of the clays and murram for dam purposes e.g. the consolidation and stability characteristics of the clay and murram respectively, the optimum moisture content for compaction in each case etc.

From such investigations, the following results were presented for certain three sites in the Kiboga District of Uganda as follows:

i. Six soil profiles were presented, two for each site (see Figures 5.4A1–C2)
ii A permeability test report was also presented (see Table 5.10).

A1

GWT – not reached

depth (m)	soil description	sample u-100 u-35	SPT Blows	Value
0	TOP SOIL			
1	Greyish brown black inorganic sandy silty			
2	clay of high plasticity (stiff)			
3		D	4/5/6	11
4	clays and clayey silt/phyllite (stiff)	U		
5				
6		D	4/5/7	12
7	clay and clayey silt/phyllite (stiff)	D	7/6/8	14
7.95	END OF BH			

KASEJJERE VALLEY DAM KYANKWANAZI SITE B H 1A

A2

GWT – not reached

depth (m)	soil description	sample u-100 u-35	SPT Blows	Value
0	TOP SOIL			
1	Grey clayey silts with some sand content			
2	(very stiff)			
		D	6/8/10	18
2.45	END OF BH			

KASEJJERE VALLEY DAM KYANKWANAZI SITE B H 1B

B1

GWT - not reached

depth (m)	soil description		sample u-100 u-35	SPT Blows	Value
0	TOP SOIL	/////////			
1	Gravelly sand with small clay content				
2	(very firm)				
3			D	13 14 14	28
4	Sand with small clay content (very firm)		U		
5				10 9	
6.45	—End of B H—		D	12	21

NAKAKABALA VALLEY DAM MASODDE SITE B H 2A

B2

GWT-not reached

depth (m)	soil description		sample u-100 u-35	Blows	Value
0	TOP SOIL	/////////			
1	Gravelly sand with small clay content				
2	(very firm)				
3			D	12 15 15	30
3.45	—End of B H				

NAKAKABALA VALLEY DAM MASODDE SITE B H 2B

C1

```
                                           GWT - Not reached
┌──────┬─────────────────────┬──────────┬──────────────┬──────────────┐
│Depth │  Soil Description   │          │   Sample     │     SPT      │
│ (m)  │                     │          │ U-100  U-35  │ Blows  value │
├──────┼─────────────────────┼──────────┼──────────────┼──────────────┤
│  0   │     TOP SOIL        │ //////// │              │              │
│  1   │ Grey-brown clay of  │          │              │              │
│      │ intermediate to high│          │              │              │
│  2   │ plasticity          │          │              │              │
│      │ (very stiff)        │          │              │      5       │
│  3   │                     │          │       D      │     10       │
│                                                             11    21 │
│  4   │ Brown gravel - sands│          │              │              │
│      │ with small clay     │          │       U      │              │
│  5   │ content             │          │              │     12       │
│                                                             14       │
│ 6·45 │ ___ END OF BH ___   │          │       D      │     15    29 │
└──────┴─────────────────────┴──────────┴──────────────┴──────────────┘
```

RWABIKUNYU VALLEY DAM BUKOMERO SITE BH 3A

C2

```
                                           GWT - Not reached
┌──────┬─────────────────────┬──────────┬──────────────┬──────────────┐
│Depth │  Soil Description   │          │   Sample     │     SPT      │
│ (m)  │                     │          │ U-100  U-35  │ Blows  Value │
├──────┼─────────────────────┼──────────┼──────────────┼──────────────┤
│  0   │     TOP SOIL        │ //////// │              │              │
│  1   │ Grey-brown inorganic│          │              │              │
│      │ clays of intermidiate│         │              │              │
│  2   │ plasticity          │          │              │              │
│      │ (very stiff)        │          │              │      6       │
│  3   │                     │          │              │      8       │
│ 3·45 │ ___ END OF BH ___   │          │       D      │     12    20 │
└──────┴─────────────────────┴──────────┴──────────────┴──────────────┘
```

RWABIKUNYU VALLEY DAM BUKOMERO SITE BH 3B

Figure 5.4: Soil Profiles and Reports of Soil Tests

From this information, the report concluded:

i. The site investigation carried out has provided information about the type of soils available at the three sites proposed for construction of dams/valley tanks.
ii. The stratigraphy is typical of tropical weathering of certain rocks leading to the formation of the type of soils described. No bedrock was struck during the site investigations.
iii. Inference into the permeability values obtained indicated that at a depth of about 4 m, the permeability values of the strata are very low according to the ratings by Terzaghi and Peck.

TABLE 5.10   PERMEABILITY TEST RESULTS ON UNDISTURBED SAMPLES

SAMPLE	COEFFICIENT OF PERMEABILITY (M/S)	RANGE OF COEFFICIENT OF PERMEABILITY (M/S) BY TERZAGHI AND PECK	
1. BH 1A (Depth 4.50–4.95 m)	5.12×10^{-9}	$> 10^{-3}$	High
		10^{-3} to 10^{-5}	Medium
2. BH 2A (Depth 4.50–4.95 m)	8.66×10^{-8}	10^{-5} to 10^{-7}	Low
		10^{-7} to 10^{-9}	Very Low
3. BH 3A (Depth 4.50–4.95 m)	3.07×10^{-8}	$< 10^{-9}$	Impervious

BH 1A = Kasejjere Dam Site
BH 2A = Nakakabala Valley Tank Site
BH 3A = Wabikunyu Valley Tank Site

The conclusions and recommendations on five other sites is even more revealing.

i. Nyarubungo Katereera Dam:

From the analysis of the four test pits dug to an average depth of 3 m, (no water encountered), as well as laboratory tests, the following conclusions and recommendations were made:
   a. The soil investigation indicates that the soil encountered in the bottom of the reservoir area is fairly impermeable clayey soil with a good water-retaining capacity to a depth of approximately 2 m. Below the clayey soil, the content of gravel fraction increases. The soil below 2 m therefore has a higher permeability.
   b. The existing valley tank had no water at the end of the dry season in September 1995. The depth of the valley tank is more than 2 m; there is therefore likely to be seepage into the lower soil layers which have a higher permeability.

It is recommended that:
   c. The existing valley tank is filled with compacted clayey soil in the course of the construction of the new dam.
   d. Suitable embankment fill material can be excavated near test pit (TP) 1. A homogeneous embankment is recommended in view of the preponderance of sandy clay materials found around TP 1, up to a depth of about 3 m. Excavated materials from the spillway will also be useful fill material.

ii. Kantaganya Engali Dam

Four TPs were also dug to an average depth of 3 m. No water was encountered.

Conclusions and Recommendations:
   a. The soil encountered in the bottom of the reservoir area – TPs 2, 3 and 4 – is rather varied, having a mixture of sandy layers and clayey layers. The top soil is clayey and has some water-retaining capacity, evidenced by the lack of water in the permeable sand/gravel layers below it. So, seepage may be expected if a dam is constructed at this location.
   b. It is recommended that a cut-off clay core is constructed to a minimum depth of 2 m below the ground level covering the flat portion of the valley bottom to reduce seepage below the dam.
   c. Suitable embankment fill materials can be excavated near TP 1.

iii. Kibutamo Valley Tank

The four TPs to a depth of about 1.8 m have similar profiles, no water was encountered. Conclusions and recommendations,
   a. The investigations indicate that the valley tank is located in an area of impermeable clayey soil with high plasticity and a very good water-retaining capacity. The existing valley tank is also holding water, though very little at the end of the dry season.
   b. It is recommended that the new tank is constructed as an extension of the existing tank and that the excavated material from the tank is used to construct a low berm across the very flat valley to help diversion into the tank.
   c. It is also recommended that the tank incorporates a filter and a well to improve the water quality.

iv. Mamba Swamp – Valley Tank:

Three TPs were dug of depths of 2.8–3.7 m. The pits indicate the presence of rock from depth 2.8 m and rock boulders from 3.7 m in one TP. This shows that the permeability of the soil below about 3 m is very high. This site is therefore unsuitable for a valley tank of 3.5–4 m depth without investing additional expenses on clay lining.

v. Kigaaga Dam:

Three TPs were dug along the dam axis and one plus an auger hole in the reservoir area; the depths vary from 2.8 m for the auger hole to 3.1 m for the TPs.
   a. The TP around the spillway shows that the soil around the depth of the spillway control section (1.5–2.5 m) is highly erodible (sandy chalk).
   b. The soil in the reservoir area to about 2.5 m depth is different classes of clay of low permeability. This is good for reservoir purposes.

Based on these reports, sites (i), (ii) and (iii) were confirmed, Mamba Swamp dropped, while the spillway of Kigaaga Dam was relocated on the other side of the valley, even when this meant more excavation and therefore greater expense!

The interpretation of these reports and the actions based on them underscores the importance of soil investigation in surface reservoir design.

Chapter VI
DESIGN OF VALLEY TANKS AND DAMS

Using the information obtained from the studies discussed in chapter V, the engineer decides on the design criteria/parameters and proceeds to design the appropriate structures for the selected sites; he then concludes the design process by preparing design drawings, which represent his overall concept of the project. The procedures for valley tanks and dams are as follows.

6.1 Design of Valley Tanks and Dams

This involves the integration of the information made available from the investigations and studies outlined in chapter V. Let us use the data for Akayanja (Table 5.7) to follow the design process. First, it is necessary to define what was intended to be done and what was possible in form of design notes, as follows:

The required storage was 6,990.5 m^3 against a catchment yield of 125,750 m^3. Because of this copious catchment yield, a bigger storage of about 12,000 m^3 was proposed to take advantage of the present development efforts to provide for the imminent higher human and livestock populations in the area. However, because of budget limitations, the project provided a 5,805 m^3 valley tank and located the site of the bigger tank both on the topographical plan and on the ground.

The basic computation in the design of a valley tank is the volume of water to be stored in the tank, which is essentially the same as the volume of excavation to be made to create the storage tank. Typical tank arrangements are shown in Figures 6.1A and B.

Figure 6.1A shows the Kanyaryeru Ministry of Works (MOW) Valley tank (Table 5.7), an arrangement in a flat terrain, where there is no defined drainage channel, and therefore requires inlet channels to guide the run-off into the tank. Figure 6.1B shows one in a defined valley, where a grassed filter strip is provided to reduce the rate of siltation of the tank. The valley in Figure 6.1B is actually a good dam site, but a tank is selected because of the low catchment yield, which is too small to justify the cost of dam; the site is Mpangamushanju in Table 5.7.

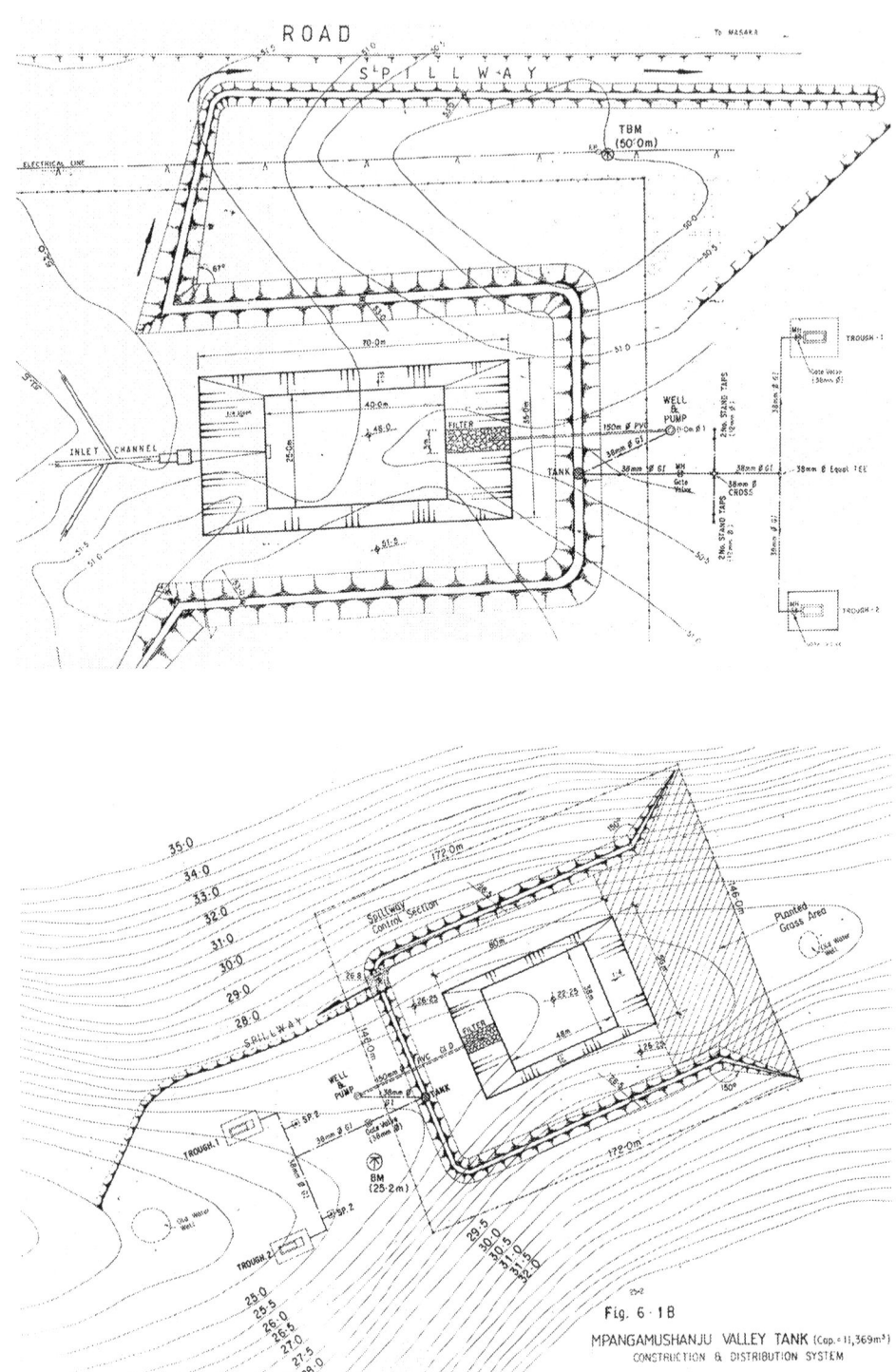

Figures 6.1A and B: Typical Valley Tank Arrangements

The capacity of a trapezoidal tank C is obtained from the relationship:

$$C = [A + B + \sqrt{AB}]D/3 \quad \ldots\ldots 3$$

where A = Area of tank bottom
B = Area of top of tank and
D (or H) = Depth of tank

When applied to the Akayanja situation, the two tanks proposed have the capacities shown in Table 6.1

TABLE 6.1   AKAYANJA (UGANDA) VALLEY TANK PARAMETERS

PARAMETER	PRESENT TANK	FUTURE TANK
Top dimensions	70 m x 35 m	80 m x 50 m
Bottom dimensions	40 m x 25 m	48 m x 38 m
Depth	3.5 m	4.2 m
Storage capacity	5,805 m^3	11,935 m^3

In practice, and depending on site conditions, the capacity could be increased by providing a spillway with a control section some convenient level above the general level around the tank.

In the case of Figure 6.1A (MOW), the spillway control section (sill) is fixed at elevation 52.0 m, which is 0.5 m above the valley tank ream. This provides additional storage of 0.5 m x surface area of the space enclosed by the perimeter bonds around the tank at elevation 51.5 m = 0.5 x 100 m x 55 m = 2,750 m^3.

Thus, the tank actually has a total storage capacity of 5805 + 2750 = 8,555 m^3, which can still be filled by the catchment yield of 58,425 m^3.

Similarly for Figure 6.1B (Mpangamushanju), with the spillway control at elevation 26.8 m, i.e. 0.55 m above the tank ream, additional storage of 7,425 m^3 (0.55 x 150 m x 90 m) is provided, giving a total storage capacity of 18,794 m^3 (11,369 + 7,425), which can be filled *only* in an above-average rainfall year, since the mean annual catchment yield is only 12,245 m^3. All the same, the design maximises the site potential for storage.

These calculations of additional storage can be refined by multiplying the average of the surface areas at the elevation of the tank ream and the water surface by the water depth above the tank ream. It is, however assumed here that the difference in these surface areas for a water depth of less than 1 m is negligible.

6.2   Design of Dams

With the results/reports of the three design tools at his disposal, i.e. the results of hydrological studies, topographical survey and geotechnical investigations, the engineer proceeds with dam design.

6.2.1 INTRODUCTION

The issues discussed hereunder relate to the principles on which the design of a dam are based, the basic concern being to design a dam and appurtenant structures that conform with the basic technical requirements of such structures at the least first (initial) costs, which would ensure that they can retain water throughout their useful lives with minimal maintenance costs and danger to lives and property.

6.2.2 PLAN OF DAM AND APPURTENANT STRUCTURES

Figure 6.2 shows the typical plan of a dam and appurtenant structures, superimposed on a topographical plan of the site. Notice that if located at any other point on the topo plan, the dam axis would be longer and construction costs would be higher for the same dam height and reservoir capacity.

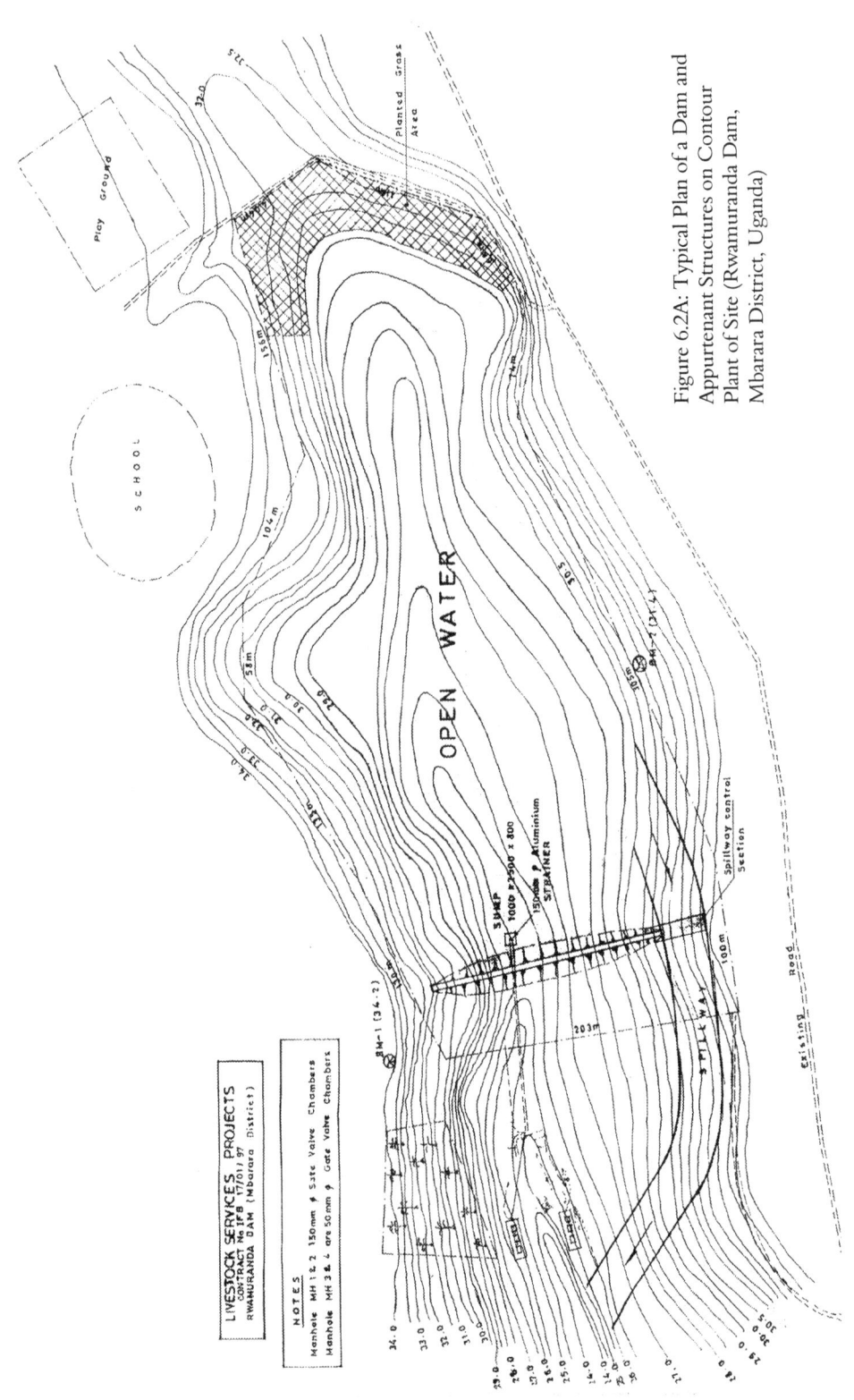

Figure 6.2A: Typical Plan of a Dam and Appurtenant Structures on Contour Plant of Site (Rwamuranda Dam, Mbarara District, Uganda)

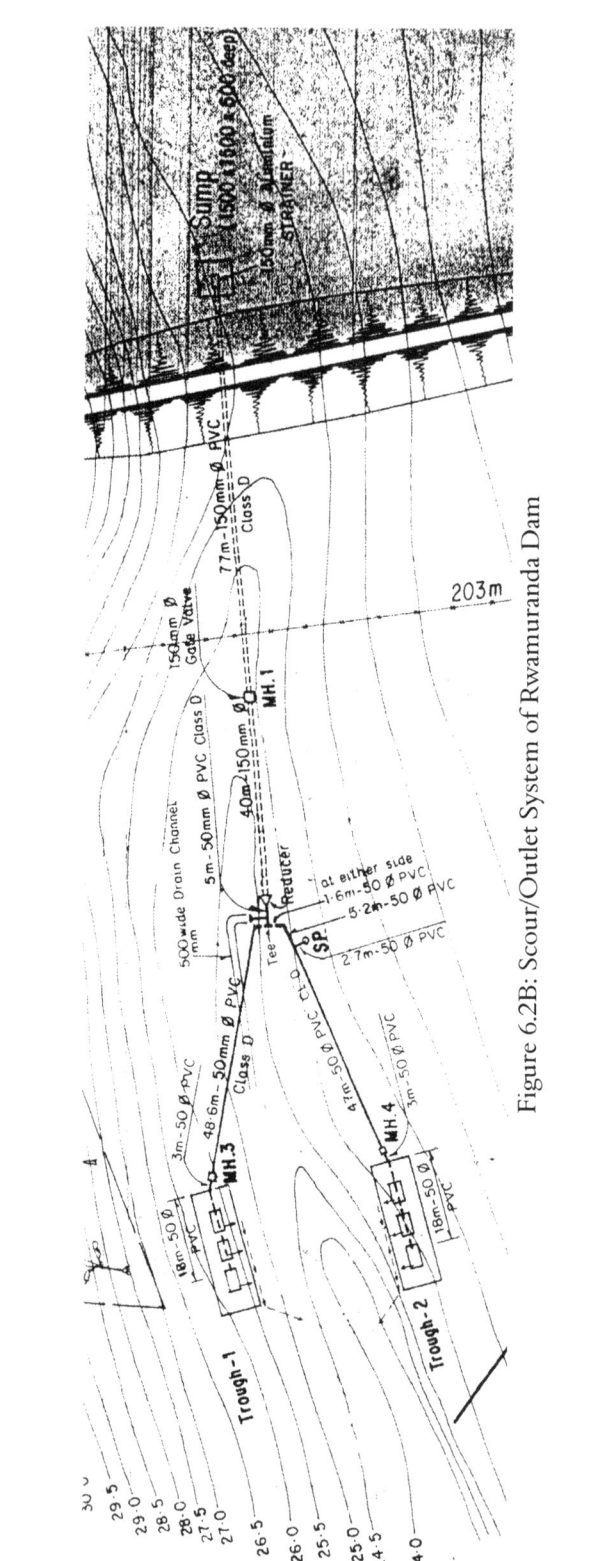

Figure 6.2B: Scour/Outlet System of Rwamuranda Dam

6.2.3 EMBANKMENT TYPE

The zoned rolled earthfill embankment type is presented here as the typical embankment to ensure stability and reduce seepage to the barest practicable minimum. The embankment is made up of a central impervious core, flanked on the upstream and downstream sides by zones of pervious materials. The requirements of a good embankment are that:

i. its slopes must be stable under all conditions of construction and reservoir operation
ii. it should not build up excessive stresses in the foundation
iii. seepage through it should be minimal and controlled
iv. it should be safe against overtopping
v. it must be stable under appropriate seismic conditions.

These factors should be taken into account in the design. The dimensions of the different zones and the maximum section for each dam should also be based on its site-specific parameters. Details are usually given in the design drawings, as in Figure 6.3.

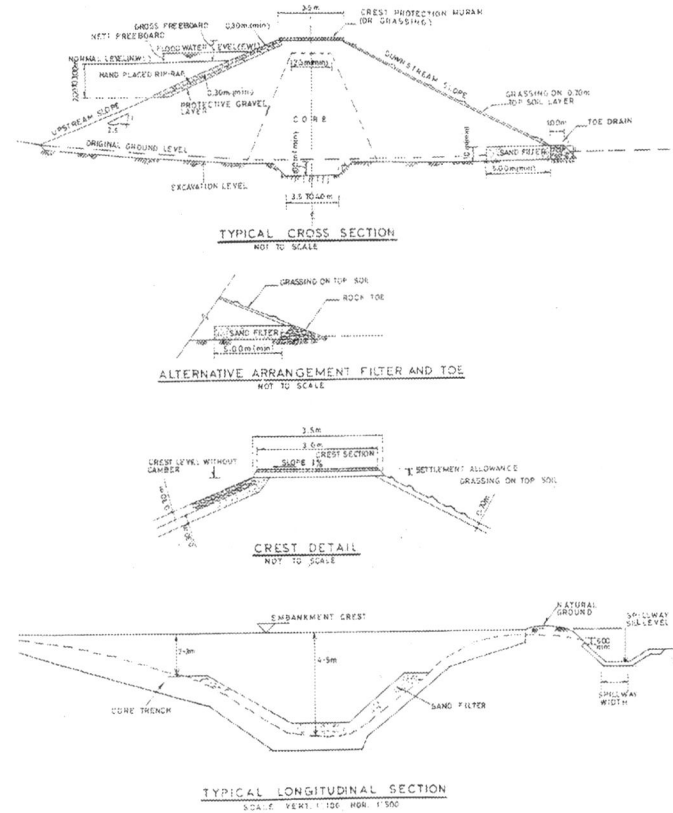

Figure 6.3: Typical Cross-Sections of a Dam

6.2.4 FOUNDATION TREATMENT

The foundation treatment for each dam should be based on the geo-technical investigation reports, ensuring that the essential foundation requirements are met, viz to provide:

i. a suitable support for the embankment under all conditions of saturation and loading, and
ii. sufficient resistance to seepage which may endanger the stability of the dam as well as lead to considerable loss of water.

The two common foundation categories generally encountered in semi-arid areas are coarse grained (sand and gravel) and fine grained (silt and clay); rock foundations are not common, though will be considered briefly for the sake of completeness (see 6.2.7). The specifications and bills of quantities give the site-specific details and the foundation treatment designed for each site condition.

6.2.5 SEEPAGE CONTROL

Seepage control is effected through the several components of the dam that are prone to it, i.e.

1. Foundation
2. Embankment
3. Drainage/Outlet structure.

The embankment type (zoned rolled earthfill) and foundation treatment are aimed essentially at dam stability and seepage control. While the central impervious clay core limits seepage through the embankment, the more pervious upstream side shell provides stability against rapid drawdown of the reservoir (which will be the case in the dry season for virtually all small dams in semi-arid regions); the downstream side shell acts as a drain to control seepage and lower the phreatic surface (upper surface of seepage or zero pressure line). Both shells enclose, support and protect the central impervious clay core. See 6.2.13 for seepage control along outlet/drainage structure.

Good impervious clay and murram (soils with fairly high clay content but more pervious than clay) are generally found along most valleys and should be taken advantage of. The compaction of the embankment fill materials should be done to 95% maximum dry density (MDD), and the materials laid in successive lifts of not more than 150 mm post-compaction thickness, preceded by watering if necessary, to attain the optimum moisture content for maximum compaction. Where homogeneous embankments are adopted, compaction should follow the same procedure.

In addition, the introduction of key trench into firm material in all cases will penetrate any soil layer of doubtful permeability within the foundation depth, thus further reducing the chances of seepage under the foundation.

6.2.6 SEEPAGE CONTROL IN DIFFERENT FOUNDATION TYPES

6.2.6.1 *Introduction*

Foundation treatments for different materials vary. It is an established fact that most earth dam accidents, including failures, result from inadequate foundation treatments. This underscores the fact that foundation treatment is very crucial in dam design and construction.

Foundation problems could be complex, requiring detailed and extensive field and laboratory investigations, which are inevitably expensive. These detailed explorations and the consequent theoretical designs are not required for small earth dams. It is usually more economical to use empirical foundation designs, being conscious of and building safety factors into the design as much as possible. But where complex foundation situations are encountered, detailed investigations/tests are inevitable. However, in consideration of costs, a change of site could be considered.

The more commonly encountered foundations are grouped into three as follows:

1. Rock foundations
2. Coarse-grained (sand & gravel) foundations and
3. Fine-grained material (silt & clay) foundations.

The required treatments for each type are summarised here; more detailed information can be found in the relevant references.

6.2.6.2. *General*

Where the foundation material is impervious and of characteristics similar to those of a compacted embankment (see Kasejjere Report, Figure 5.4 1A and B and Table 5.10), the minimum treatment required for any foundation should be adopted, i.e. stripping to remove topsoil with organic material and other unsuitable materials. This can be done by open excavation to a depth of about 0.5 m in sparsely vegetated semi-arid areas. Where the overburden is shallow (most semi-arid areas are underlain by the basement complex), the whole overburden can be stripped to bedrock. These foundations serve as key trenches, (keying the fill into the parent foundation materials). A key trench bottom width of 3.5–5 m is generally adopted.

It is also common, in fact, more often than not, to find a combination of the three foundation types on a dam axis – the valley bottom (stream bed) usually of sand-gravel material, exposed rock abutments on the steep portions and clay or silt on the gentle. It may therefore be necessary to apply different treatments to the different situations encountered.

6.2.7 FOUNDATION TREATMENT METHODS

6.2.7.1 *Rock Foundations*

This is generally considered the more competent foundation type, which usually does not present problems for small dams. Even foundations of weaker rock are preferred to soil foundations. The choice of a rock foundation is justified where:

1. The rock mass is generally homogeneous and competent throughout the foundation zones that would be affected and the reservoir, but
2. They must be adequately investigated to ensure that they are adequately competent (if in doubt, consult an experienced earth dam designer.)
3. Foundation rock surfaces against which fill is to be placed, must be properly treated to ensure that fractures, fault zones, steep faces, rough areas and weathered zones do not lead to seepage and piping in the interface between foundation and fill. Treatment of deficient areas is critical for areas beneath the impervious core and the filter and drainage zones immediately downstream of the impervious zone.

Whether or not a foundation should be grouted should be determined by examining the site geology and by analysing the water losses through foundation exploration holes. Much experience is required in handling the problem of rocky foundation treatment because each foundation is unique. There may also be more economical (cheaper) or effective methods of controlling seepage or leakage than by grouting.

Grouting is a process of injecting under pressure a fluid sealing material into the underlying formations through specially drilled holes to seal off or fill joints, fractures, fissures, bedding planes, cavities or other openings in rocks. The services of an experienced specialist in the field should be employed once it is suspected through preliminary investigations that grouting may be necessary.

Filters and drains are the most important features for collecting and controlling seepage through rock foundations. Even when a rock foundation is grouted and cut-off provided, appropriate filters and drains are still needed to collect seepage and reduce uplift pressures in the areas downstream of the impervious zone.

6.2.7.2 Sand and Gravel Foundations

Generally, sand and gravel foundations have adequate strength to sufficiently support the loads induced by the embankment and reservoir; this must be properly investigated however.

Two problems associated with this foundation are: underseepage and seepage forces.

Among the methods used to treat sand and gravel foundations are:

- Cut-off trenches
- upstream blankets
- pressure relief wells
- filters and drains and

- extended downstream pervious zone (i.e. base width beyond the core of impervious material).

Investigations will usually guide one on the type of treatment required. Adequate measures must then be built into the design to ensure safety against failures occasioned by piping.

6.2.7.3 Silt and Clay Foundations (Fine-grained Soils)

The main problem with this foundation is stability because of the low-bearing capacity and the tendency to disintegrate when saturated, leading to considerable settling. Treatment should therefore be based on: soil type, location of water table and soil density.

If these materials occur in pockets, it may be sufficient to remove the very soft materials, but in most cases, the entire foundation or the bulk of it is made up of soft materials.

Again, the investigation reports will usually guide the choice of treatment to apply. Among the treatments commonly used are:

1. Use of horizontal and inclined filter drainage blanket
2. Recommended slopes of stabilizing fills for dams on saturated silt and clay foundations.

These and other methods are covered extensively in USBI, BR, *Design of Small Dams* and many others.

TABLE 6.3 RECOMMENDED SLOPES FOR SMALL ZONED EARTHFILL DAMS ON STABLE FOUNDATIONS

DAM TYPE	PURPOSE	SUBJECT TO RAPID DRAWDOWN	SHELL MATERIAL CLASSIFICATION	CORE MATERIAL CLASSIFICATION	UPSTREAM SLOPE	DOWNSTREAM SLOPE
Zoned with min. core	Any	Not critical	GW, GP, SW or SP (gravelly)	GC, GM, SC, SM, CL, ML, CH or MH	2:1	2:1
Zoned with max. core	Detention or Storage	No	GW, GP SW or sp (gravelly)	GC, GM, SC, SM, CL, ML, CH, MH	2:1 2.25:1 2.5:1 3:1	2:1 2.25:1 2.5:1 3:1
Zoned with max. Core	Storage	Yes	GW, GP, SW or SP (gravelly)	GC, GM, SC, SM, CL, ML, CH, MH	2.5:1 2.5:1 3:1 3.5:1	2:1 2.25:1 2.5:1 3:1

SOURCE: USDI, BUREAU OF RECLAMATION, *DESIGN OF SMALL DAMS*, 3RD ED., 1987 (P 252).

Min. Core = core width of 3–35 m
Max. Core = core width of 3–130 m
Rapid drawdown = 150 mm or more per day after prolonged storage at high reservoir levels.

6.2.8 SELECTION OF EMBANKMENT SLOPES

i. The choice of embankment slopes is a function of the soil materials to be used for the dam shell (wall). Based on the soil information collected during site selection, the soils should be classified according to the Unified American Association of State Highway and Transportation Organization (AASHTO) and the United States Department of Agriculture (USDA) systems of soil classification according to the United States Soil Conservation Service (USSCS), (Table 6.3). This could be confirmed by the prescribed geotechnical investigations.

For zoned rolled earthfill embankments provided with a centrally located impervious core on adequately strong foundations, upstream and downstream slopes of 2:1 are suitable for dams not more than 15 m above the lowest point in the stream bed, even if subject to rapid drawdown[11]. The shell materials (murram) are generally GP, i.e. poorly graded gravel (some well graded – GW), while the core is usually clay (CL), fat clay (CH) and, in some cases, clayey sand (SC). The core and murram have permeability values of poor to practically impermeable, with coefficient of permeability ranging from 1×10^{-2} for murram to $1 \times 10^{-4} - 10^{-8}$ for clay – (compare with Table 5.10).

ii. In addition to stability considerations in side slope selection, cost is another equally important factor. For a given dam height H, the flatter the slope, the more costly the dam, since earthworks volume increases. Table 6.4 gives a cost comparison for Kenwa Dam in Mbarara District, Uganda, using a constant downstream slope of 2:1 against upstream slopes of 2.5:1 and 3:1.

TABLE 6.4 COSTS OF EARTH EMBANKMENTS AS AFFECTED BY UPSTREAM SLOPE

SLOPE	UPSTREAM SHELL		SLOPE TRIMMING		GRASSING		TOTAL
	M^3	SH.M.	M^2	SH.M.	M^2	SH.M.	SH.M
2:1	7614	45.68	1723	7.75	1723	1.72	55.15
2.5:1	8483	50.90	2415	10.87	2415	2.42	64.19
3:1	9334	56.00	2903	13.06	2903	2.90	71.96

Additional cost for upstream slope of 2.5:1 = Sh 9.04 million and Sh.16.81m for 3:1. Details of the derivation of these costs are given in Appendix 2.

Table 6.4 shows that if an upstream slope of 3:1 were adopted, the cost of Kenwa Dam would go up by at least Sh.16.5 million, using the contractor's rates for the dam as awarded. One could adopt an upstream slope of either 2.5:1 or 3:1 if the soil classes of the shell materials make it expedient (Table 6.3). But in this case, adopting a gentler slope than 2:1 (i.e. either 2.5:1 or 3:1)

---

[11] USBR, *Design of Small Dams*, A Water Resources Technical Publication, 1987 ed., p 252; rapid drawdown is 150 mm or more per day after prolonged storage at high reservoir levels.

would increase the cost of the dam in question without adding quality to its stability or functionality.

6.2.9 FREEBOARD

Freeboard (referred to as gross freeboard in Figure 6.3) is the vertical distance between the embankment crest and the water level in the full reservoir; it is provided to prevent overtopping of the embankment by wave action and the occurrence of an inflow flood greater than the design inflow, all leading to a rise in the maximum water level.

Based on this premise, emergency spillways are recommended for these areas; they are usually ungated open side channels with control crests located at the design normal water level (NWL). Essentially they are overflow rectangular broad-crested weirs with the following advantages:

i. ease of construction
ii. automatic and trouble-free operation
iii. ability to function without an attendant, and
iv. low maintenance cost.

To function effectively in preventing the embankment from being overtopped, they are designed to offer resistance to erosion, greater than does the embankment itself; they are to be cut through ridges and embankment abutments, their discharge outlets are also located far from the dam toe to preclude damage to the dam and the outlet structures. Care must be taken in the course of construction to ensure that the control section is at 0.0 level to make it effective.

The channels are designed to discharge the floods expected from a maximum daily rainfall of a return period of say 1:20 years as is usual for small catchments in semi-arid areas.

The computation of the inflow design flood is based on the 'rational approach' for the design of Emergency Spillways (FAO 1975).[12]

The peak discharge or maximum inflow is given as:

$$Q_p = CIA/360 \ldots 4$$

where Q_p = Peak rate of flow (m³/hr)
C = Constant catchment coefficient (Table 6.5)
I = Maximum rainfall intensity (mm/hr)
A = Catchment (drainage) area (km²)

As an example, based on a rainfall intensity-duration relationship study carried

---

[12] FAO, 1974, Effective Rainfall, Irrigation and Drainage Paper 25.

out by Hydromet (1974), at Ntusi,[13] Uganda, the intensity (I) of long rainfalls which generate run-off that could lead to floods of recognisable magnitude (duration longer than 6 hours) is 10 mm/hr, obtainable from the relationship:

$$I = 53.6/(t + 0.65)1.042 \text{ (mm/hr)} \ldots 5$$
where I = rainfall intensity (mm/hr)
t = duration (hrs.)

Thus, the spillway discharge for a few dams in Uganda's cattle corridor, derived from equation 4, using C = 0.35 (i.e. pasture land with average infiltration rates) selected from Table 6.5 are given in Table 6.6, along with other parameters.

Surcharge (or net freeboard) is the maximum allowable rise in water level above the spillway control section. This is usually calculated by routing a standard flood of known parameters through the full reservoir. In the absence of such flood data, a 400 mm surcharge is assumed, based on experience with small dams. Given a gross freeboard of 1 m, the maximum water level would be 600 mm below the embankment crest (see also 5.6).

To guard against seepage in the event of surcharge coming into effect, the core in a zoned dam should go up to 0.6 m above the spillway level while the remaining 0.4 m is made of the same fill material as for the upstream and downstream shells for protection of the core. A 150 mm camber of lateritic/gravel materials, sloped towards the reservoir for drainage purposes, should be applied to protect the crest; for road purposes, the gravelly layer should be increased to 300 mm especially where the dam crest is likely to be used as motorable road (see crest details in Figure 6.3).

6.2.10 SLOPE PROTECTION AND SURFACE DRAINAGE

Ideally, a 300 mm thick dumped riprap of hard core, dense (specific gravity over 2.5), durable and able to withstand long exposure to weathering, like granite, should be used to protect the upstream slope, from the dam crest to about 1 m below the NWL. However, because this could be expensive for a small dam, with small reservoir fetch, grassing could be adopted as a second best, more so if there are no data on wave action. For maximum effectiveness, this grass should be well established on a 200 mm topsoil bedding before the reservoir fills.

Similarly, the downstream slope as well as the control and discharge sections of the spillway could be protected with Grassing.

Where cost is a constraint however, the control section alone can be grassed as part of the construction contracts; in such cases, the water users can be educated on how to encourage grass to grow in the approach and discharge channels to minimise the development of gullies.

A drainage gutter (toe drain, see Figure 6.3) filled with cobbles could also

---

[13]Ntusi Meteorological Station is situated in Sembabule District, in the same hydro-meteorological regime as most arid and semi-arid areas; the derived relationship which has been recommended for estimating flood flows for Spillway Design has been adopted because the data are considered applicable here.

be provided along the point of contact between the downstream slope and the ground and lead into the stream bed behind the dam to prevent a gradual erosion of the embankment and to serve as filter to keep soil particles in place so that any incidental seepage may not develop into piping.

6.2.11 SPILLWAY DESIGN DISCHARGE (SEE ALSO 5.6)

Spillways are usually designed to cope with the discharge of some known flood magnitude, using hydrological analysis of any available rainfall/run-off data.

On the other hand, emergency spillways are provided primarily to avoid an overtopping of the embankment in the event of an emergency condition. But in arid and semi-arid areas, the desire is to store as much of the available run-off as possible. This desire serves as a driving force to maximise the physical potentials of the sites, provides additional storage space for run-off in years of above-average rains and reduces the Spillways essentially to emergency spillways which are not expected to function under normal reservoir operations.

It is important to note that additional safety for the dam is provided by installing drainage pipes fitted with gate valves, which can be opened by the beneficiaries themselves in times of emergency to complement the spillway in quickly lowering the reservoir level and thus reduce the chances of the embankment being overtopped in case of an unprecedented flood.

TABLE 6.5 CATCHMENT COEFFICIENT C FOR FLOOD ESTIMATION

SOIL TYPE	WATERSHED COVER C		
	CULTIVATED	PASTURE	WOODLAND
With above-average infiltration rates; usually sandy or gravelly	0.20	0.15	0.10
With average infiltration rates, no clay pans; loam and similar soils	0.40	0.35	0.30
With below-average infiltration rates; heavy clay soils or soils with clay pan near the surface or shallow soils above impervious rock	0.50	0.45	0.40

SOURCE: CHOW, VENTE, *HANDBOOK OF APPLIED HYDROLOGY*, MCGRAW-HILL INC., 1964

TABLE 6.6 SITE-SPECIFIC DESIGN PARAMETERS

DISTRICT/SITE	STRUCTURE TYPE	RESERVOIR CAPACITY AT NWL ($\times 10^3$ M^3)	EMBANKMENT LENGTH (M)	EMBANKMENT MAX. HT. (M)	SP/WAY PEAK DISCHARGE Q_P (M^3/SEC.)
MBARARA					
1. Kyera	New Dam	183,000	120	4.0	12.2
2. Kenwa	New Dam	224,000	198	6.2	24
3. Kishangura	Dam Rehab.	212,500[1]	175	4.5	
NTUNGAMO					
4. Kigaaga	New Dam	203,000	235	6.0	29.2
SEMBABULE					
5. Rwamakara	Dam Rehab.	879,000[2]	170	7.6	35
MUBENDE					
6. Dyangoma	Dam Rehab.	145,000[3]	123	7.6	9.6
KIBOGA					
7. Kasejjere	New Dam	290,000	405	5.2	10.5
NAKASONGOLA					
8. Migeera	New Dam	206,000	317	? 5.75	51.3
9. Wabale	New Dam	185,000	402	6.25	28

Reservoir Storage of Rehab. Dams

} Storage before rehabilitation [1] = 156,250   [2] = 355,000   [3] = 8,500
} Additional storage as a result of rehabilitation = 56,250   = 524,000   = 136,500
} TOTAL after rehab. = 212,500   = 879,000   = 145,000

$$Q_P = CIA$$

where C = catchment coeff.
I = rain intensity
A = catchment area (km^2)

6.2.12 ABATEMENT OF SEDIMENTATION

Sediment deposition in dam reservoirs (siltation or sedimentation) is the result of soil erosion by water in the catchment area. Soil erosion is usually severe where the soils dislodged by raindrop impact and puddling through human/livestock activities get moved downslope by run-off, which ultimately develops under two conditions:

a. if precipitation rate exceeds the infiltration rate of the soil, and
b. if there is a slope to activate gravity flow (movement) of the accumulated 'surplus' surface water.

The surplus water ultimately finds its way into the reservoir with its load of soil particles and gradually silts up the reservoir. These erosion-causing conditions are inherent in most climates where water has to be made available to make up for the inherent shortfall; therefore, specific soil and water conservation measures must be taken or built into the dam/reservoir design to minimise siltation, which is likely to occur, sooner or later. These include:

i. Cultivation should not be allowed in the catchment area, and if it has to be allowed, soil conservation measures must be taken, e.g. contour ploughing, terracing etc.
ii. Avoiding bush burning and overgrazing.
iii. Introduction of fenced filter strips of grass of some 30 m wide, immediately upstream of the reservoir along and around the main valley (see figures 6.1B and 6.2).
iv. Limiting reservoir clearing to the removal of big trees and shrubs and doing it early in the construction programme to encourage the regrowth of grass and shallow-rooted vegetation that would further filter the run-off as it fills up the reservoir.

Beneficiaries should be actively involved in the planning and construction so that they get convinced of the need to take these measures, most of which are post-construction activities. They should also be advised to arrest promptly, the invasion of aquatic plants that are known to have reduced the useful lives of similar water-retaining facilities in similar circumstances.

For more in-depth discussion of sedimentation, see Water Resources Development and Environment Management – chapter IX.

6.2.13 DRAINAGE AND OUTLET STRUCTURES

To scour is to clear out by flushing through with water or by natural action of water. A scour/drainage structure is therefore one designed to drain or flush out sediment from the reservoir. For small dams, the outlet structure is usually combined with the scour system and it consists of a pipe/conduit system leading from the deepest part of the reservoir to a point well downstream of the dam, usually with one (but better with two) valves, controlling the flow through the pipe. The outlet/scour pipe and the connected washout valve perform the following functions:

i. regulating reservoir level during times of flood or in the event of damage, to empty the reservoir for repairs
ii. for flushing the system to control siltation
iii. delivery of water by gravity to downstream usage for irrigation, livestock watering and human consumption.

They are generally designed to release water at specific rates – dictated by downstream needs, flood control regulation, storage considerations, power generation and legal requirements.

For small dams, the scour/outlet structure usually consists of a single 150 mm pipe (or sometimes 200 mm) connected to an aluminium strainer within the sump that is located in the reservoir; the pipe then runs through the embankment. It has one or two main control valves downstream (depending on whether one or two are installed for cost consideration purposes), after which it reduces to a 50 mm pipe for supply to the troughs and tap stands with a 150/50 mm reducer tee. The 150 mm end serves as a drain and is fitted with a threaded plug for ease of removal under emergency conditions. The same end can discharge into as irrigation supply canal in a multi-purpose scheme (see Figures 6.2B & 6.4A, B and C).

A. Sump with embedded scour/outlet pipe and strainer

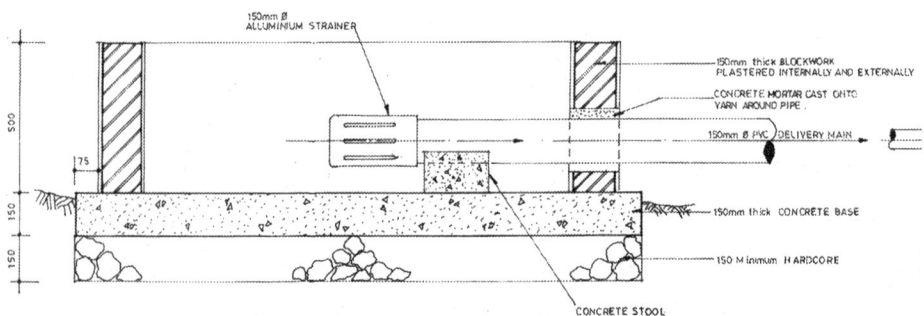

B. Gate valve protected in a manhole

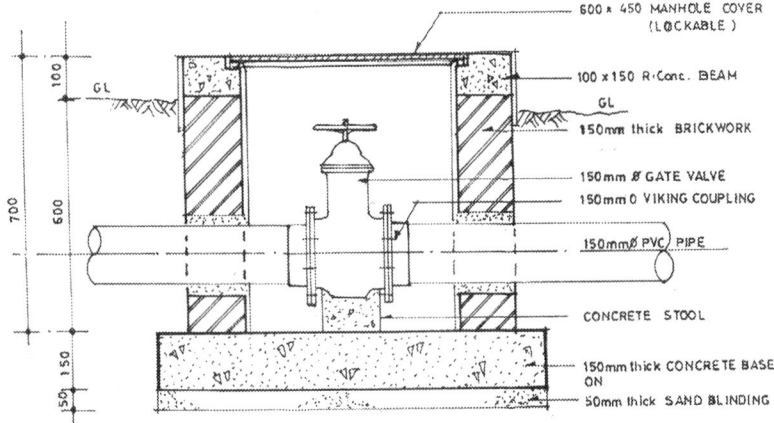

C. Scour/outlet pipe end

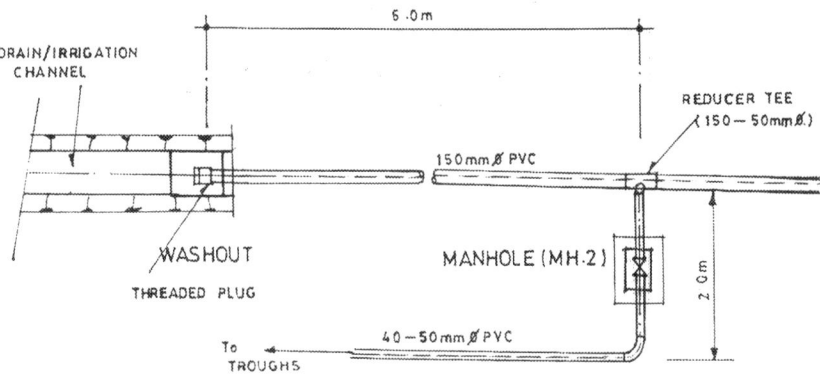

Figures 6.4: Parts of a Scour/Outlet Structure

Some General Notes on Outlet/Scour Structures follow:
1. The safety of rockfill and earthfill dams depends to a large extent on the stability of the spillway and outlet structures, especially when conduits pass through the embankment.
2. A closed conduit waterway could be an in situ cue-and-cover culvert or conduit, a precast or prefab pipe or a tunnel bored through the abutment
3. Waterways for spillways are most often free flowing, whereas those for outlet works may either flow full under pressure or partly full
4. Where all or part of a conduit is under internal pressure from the reservoir head, any leakage or failure may gradually be enlarged until partial or total failure results

5. Seepage is also possible (in fact common) along the contact surfaces between the conduit and earthfill; this can result in serious damage.
6. There is also the danger of the structural collapse of the conduit, which will almost certainly result in the collapse of the earthfill dam.

These facts emphasize the importance of using durable materials, conservative design procedures, proper design details and construction methods which would ensure safe structures.

Some guidelines on the design and installation of conduit-based (pipe) outlet structures are listed below:

i. The pipe must not be laid through the fill part of the embankment, but in a stretch excavated into the solid ground, either below (preferably) the dam foundation or to one side of the dam – to avoid the effect of differential settling of the fill on the pipe.

ii. When pre-cast pipes, including PVC, are used, the methods of bedding and backfilling the pipe should as far as possible preclude uneven settlement and ensure uniform distribution of load on the foundation. When backfilling near these structures, extreme care should be taken to ensure tight contact between the fill and the conduit surface, and to obtain the proper densities of the earthfill material – not only to prevent seepage but also to ensure that the fill develops a lateral restraint on the structure, which will prevent excessive stresses in the conduit shell.

iii. The foundation preparation and compaction around conduits must be equivalent to the foundation preparation for the dam, and compaction of the impervious earthfill.

iv. Cut-off collars increase the length of percolation path along the pipe by 20–30%, thus slowing down the rate of seepage. Hence, to minimise seepage along the pipe:

- install a puddle flange (or cut-off collar), a collar of metal or reinforced concrete, for pipe diameters $\geq$ 250 mm, projecting for about 300 mm all around it at about 1.5 m intervals (i.e. about 10 times their projection), through the pipe length within the impervious zone, i.e. the core, see Figure 6.5). To avoid concentrated stresses in the pipe, the collar should not fit tightly to the pipe; the thin space between the pipe and the collar should be filled with watertight fillers like:
    - graphite-coated paper to permit slight movement or
    - pre-moulded bituminous fillers where greater movement is anticipated
    - for small diameter pipes, $\leq$ 250 mm, graphite coated paper will suffice

- the base of the pipe section housing the flange should be concreted and possibly the entire length of the pipe that runs through the impervious zone of the embankment.
- in two or more places, the concrete thickness should be increased to fill the pipe trench to form a key with the surrounding ground; at the same points, concrete must be added above the ground for the same reason
- pipe joints must be made watertight to prevent seepage into the surrounding embankment; joints of *in situ* concrete conduits must be sealed with waterstops and rubbergasketed joints for precast concrete pipes.

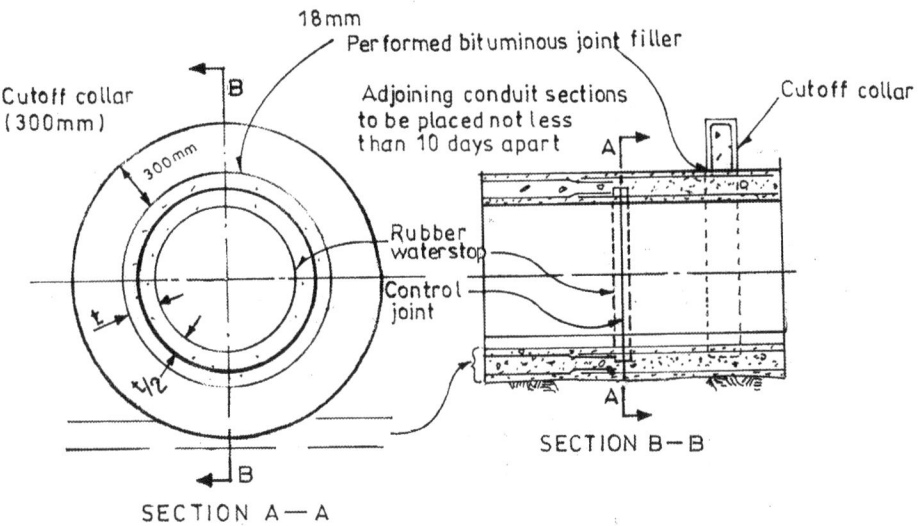

Figure 6.5: Puddle Flange (or Cutoff Collar) Around a Scour/Outlet Pipe

v. The supporting strength of precast concrete pipe under superimposed loads (e.g. under a dam) is highly dependent upon the bedding angle provided for the pipe during installation. Bedding angle is the angle formed by the arch of the pipe that is in firm complete contact with the material underneath the pipe and across which superimposed loads are transmitted from the pipe wall to the material below. One of the purposes of providing a concrete base is to provide a 90^0 bedding angle (USDI, Bureau of Reclamation).

Therefore, when the outlet consists of precast reinforced concrete, it should be set in carefully on a good foundation of reinforced concrete and well bedded in the concrete as shown in Figure 6.6.

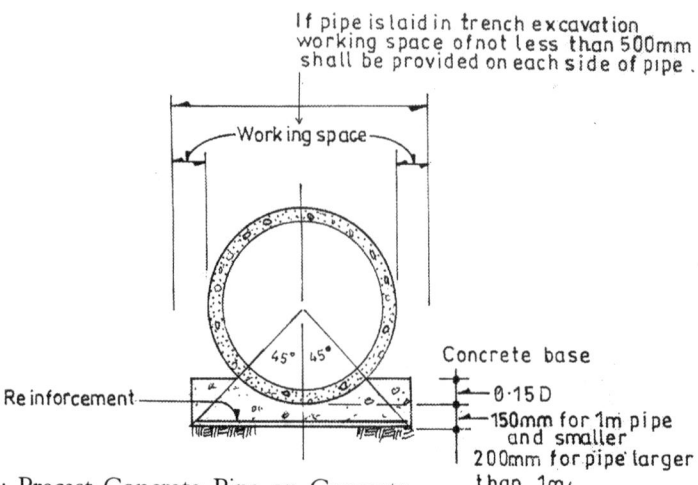

Figure 6.6: Precast Concrete Pipe on Concrete Base for Conduit Under an Embankment

This design has the following advantages:
a. distributes conduit load on the foundation,
b. precludes uncompacted zones under void spaces under the pipe, which could induce leakage along the under surface of the structure – void spaces or inadequate compaction of impervious materials at the inverts of pipes, have caused numerous failures of small earth dams.
c. prevents percolation along the underside of the pipe, where tightly compacted earth bedding is difficult to obtain. For this reason, it is advisable to place the concrete base concurrently with the pipe or after the pipe is in position.
d. provides a 90° 'bedding angle' for the precast concrete pipe.

vi. The importance of 90° bedding angle is that a pipe laid on a flat horizontal surface with line bearing on the bottom, i.e. 0° bedding angle, can support only half the load the same pipe can support when installed with a 90° bedding angle.

Valves

vii. The main valve must be well bedded in concrete (Figure 6.4B). It must be fixed on the pipe in an always accessible position, whether the dam is full or empty and even when the spillway is spilling.

viii. It is possible that a valve gets jammed with a piece of wood or stone when left fully or partly open for a long time, as in a flood, and could no longer close. Such a valve cannot be repaired without and until the reservoir is empty. For this reason, it is advisable to install a second valve at a different point along the scour/outlet pipe, so that if one jams, the water may be cut off with the second until it is convenient to clear the obstruction. Where a second valve cannot be installed, as for cost reasons, a monk can be fixed at the end of the scour pipe as a substitute (Figure 6.7).

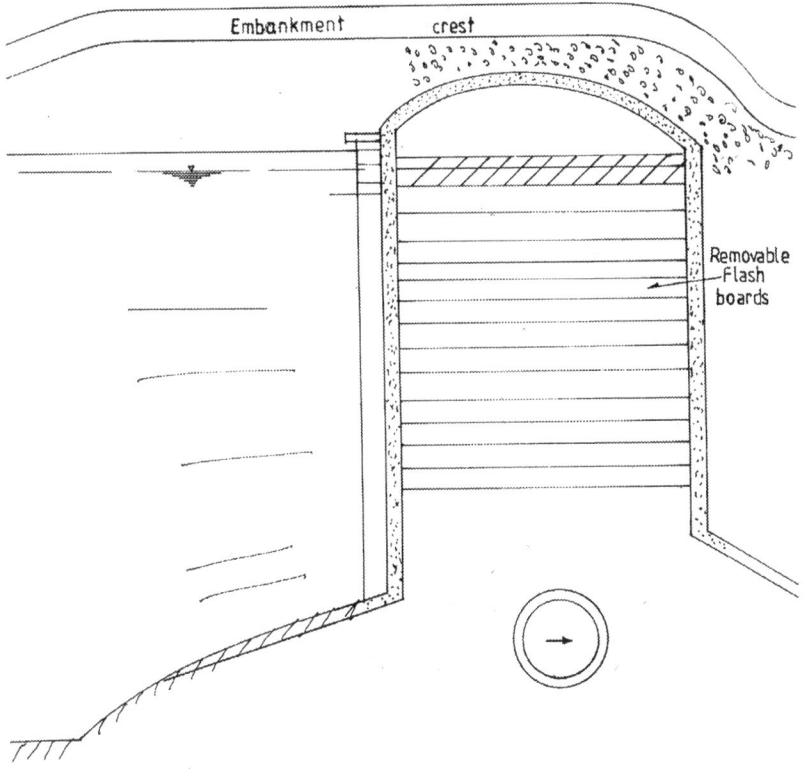

Figure 6.7: Monk

Reservoir Operation

ix. Scouring should start as soon as spilling begins
x. After full scouring for a while (i.e. valve fully opened to clear away silt immediately around the outlet), the valve should be partly closed and wastage limited to spillage so that as much water as possible is retained in the reservoir.
xi. Scouring should be kept running for much of the rainy season – as much as possible – because:
   a. it counteracts silting
   b. circulates water in the reservoir
   c. keeps the reservoir water level within safe limits at a time when a sudden storm could cause a dangerously high water level.

Materials for Conduits

Replacing conduits through either a rock-fill or an earthfill dam is usually difficult and expensive. But such an operation can be avoided by the use of durable materials, such as:

i. steel pipe enclosed in concrete
ii. *in situ* reinforced concrete conduit
iii. precast concrete pipe
iv. polyvinyl Chloride (PVC).

Until recently, (i)–(iii) above were very commonly used. But more recently, PVC has become more popular.

6.2.14 THE USE OF PVC PIPES

Because of its durability and freedom from corrosion as well as its conveyance efficiency over other materials, PVC pipe has become the obvious choice in many recent schemes. Its flexibility also reduces the chances of breakage/cracking under the load of the fill above it, even in the event of differential settling.[14]

The ideal size for small discharges is 150 mm. Pressure is minimal whether the pipe flows full or partially full, because the controlling point is at the outlet. Bernoulli Theorem of continuity also applies to both full and partially full flows for the same reason, i.e.

$$H_T = h_L + h_v \quad \ldots.6$$

where H_T = total head needed to overcome the various head losses (friction) to produce discharge
h_L = cumulative losses of the system and
h_v = velocity head at the gate valve

For practical purposes, the ratio of total head to pipe diameter (H/D) exceeds 1.5, which is the minimum gradient required to generate sufficient flows for the troughs and stand taps within a short time. The higher this ratio, the greater the flow generated.

Structural Design Considerations

Properties of PVC Pipes

CLASS	WALL THICKNESS (MM)	MAX WORKING PRESSURE (BARS)
B	4.3–5.00	6
C	6.3–7.2	9
D	8.3–9.4	10

Nominal outer diameter in all cases is 160 mm (actual – 160-160.5)
For a pipe thickness of 8.85 mm {(8.3+9.4)/2}, inner diameter = 160–8.85 = 151, practically 150 mm).

---

[14]Marston Theory of Embankment Pressure on Precast Conduits under relatively low fills (i.e. < 15 m).

The pipe is able to withstand a total internal hydrostatic pressure of:
$$P = 2yt/d\text{-}t \ldots 7$$
where y = yield stress (MPa), approximately 42 MPa (i.e. 420 m head of water)
t = minimum measured wall thickness (mm), and
d = measured outside diameter

For class D pipe, d = 160 mm, t = 0.0083 m, P = 46 m, acting against the full load of fill materials (dry or wet). Therefore, class 'D' pipe which accommodates a working stress of 12 bars i.e. 12.19806 x 10^6 kg force will certainly withstand the resultant load under dry conditions as summarised below (Marston theory).

Marston Theory/Differential Settling
Under this theory, the vertical load on a conduit is a combination of the weight of the fill directly above it plus the differential frictional forces acting either upwards or downwards from the adjacent fill. When the adjacent fill settles more than the overlying fill, downward frictional forces are induced; this increases the resultant load on the conduit. Therefore, to avoid downward frictional forces, the pip trench should be cut within a well-compacted layer so that the adjacent fill areas of the pipe are pre-compacted to specification before the pipe is installed. Better still, where the trench is cut in undisturbed firm ground, the adjacent soil is practically rigid and differential settling is virtually none existent.

On the other hand, a greater settlement immediately above the pipe results in an arching action which reduces the load on the pipe. This is likely to happen in most cases because the presence of the pipe would encourage slightly less compaction of the fill immediately on top of it in a bid to avoid breaking the pipe in the course of compaction: the pipe will therefore not receive the full weight of the fill on top of it because of arching, which comes into effect if and when the backfill starts to settle.

Shell materials (well-drained soils) have an average unit weight of 1664–1920 kg/m³ (i.e. classes GW, GP, SW and SP, according to US soil classification, Table 6.3). Similarly, the core materials (clay) have an average unit weight of 1600 kg/m³, giving a load of 216 kg over a 1 m stretch of pipe.

Even if one applies the weight of the heavier shell material over a 1 m length of the pipe installed at say 6 m below the dam crest, this gives a load of 1728 kg. Thus, an outlet pipe length of 30 m embedded within the embankment at a depth of 6 m, will experience a total load of 28,994 kg. Even when the reservoir is full and the embankment gets wet, the total load is still very small, compared with the working stress of 12 bars (120 m head of water). All the same, the effect of arching discussed above will reduce the net effect on the pipe.

Effect of Hydrostatic Pressure

Because the control valve is located at the end of the pipe, the pipe is subjected to a hydrostatic pressure, equal to the full reservoir head once the reservoir is filled and the valve opened to fill the pipeline. When applied to the deepest reservoir of 5 m, the maximum hydrostatic pressure is 5 m head of water (5,000 kg force). This pressure reduces the net effect of the embankment fill on the pipe by its value i.e. 5000 kg force (= 5 bars = 5 m head of water = 0.5 MPa).

6.2.15 ENVIRONMENTAL PROTECTION

Most of mankind's attempts to improve on nature are an intervention, which affects the natural environment, positively and or negatively; such is water resources development. The sudden presence of a relatively large body of surface water in an otherwise dry environment is bound to influence the environment and the entire ecological system. Positive impacts are desirable, in fact, they constitute the objectives of any scheme. Unfortunately, negative impacts tend to go hand-in-hand with positive impacts! To minimise the likely adverse effect of this development on the environment, the following are some of the steps to take.

i. Fencing of the embankment and reservoir areas with barbed wire on seasoned and treated timber to be supplemented with hedges by the beneficiaries, using drought-resistant shrubs common in the area. This would keep human beings and animals, especially livestock out of the fenced area and thus reduce the chances of induced erosion and water pollution.

ii. Grassing all topsoiled excavated heaps and providing a grassed strip of about 30 m upstream of the reservoir along the main drainage channel, using vetiver grass, paspallum or bahama (cynodon dactylon) – which establish quickly and are drought resistant – they serve as silt trap and slow down the rate of siltation.

iii. Adoption of soil conservation measures in the catchment areas by the beneficiaries – e.g. avoiding over-grazing and bush burning. Contour farming should be adopted if the catchment area has to be cultivated at all.

These measures are intended to prevent soil erosion upstream of the reservoir as a means of keeping low the rate of siltation of the reservoir and to preserve the embankment and spillway to prolong the life of the entire storage infrastructure.

For further discussion on this subject, see chapter IX – Water Resources Development and Environment Management.

6.3 Dam Rehabilitation

6.3.1 INTRODUCTION

Dam rehabilitation does not require area/capacity curves to determine its reservoir capacity since it is the existing features that have to be improved upon, based on the present physical features. Besides, in most cases, these dams have water in their reservoirs at the time of selection and survey; it may not be possible therefore to determine the parameters of such reservoirs precisely.

In most cases, dam rehabilitation involves embankment repair/reshaping, spillway repair and slope reshaping. To increase the reservoir capacity, it is commonly believed that reservoir desilting is the only way. Experience has shown that this is not always so however. The following case study speaks for itself.

6.3.2 KISHANGURA DAM – A CASE STUDY

Site Inspection Report

Kishangura Dam is in Nyabushozi County, Mbarara District, Uganda. At the time of inspection (August 1995), the entire site, i.e. the dam and reservoir, was very bushy; the embankment showed signs of very poor maintenance; there was some water in the reservoir but it was not possible to determine the reservoir depth. For purposes of rehabilitation, the following works were listed, subject to confirmation after topographical survey and further study.

- desilting
- installation of a water delivery system
- fencing
- stopping seepage if any.
- de-vegetation of embankment and reservoir

(See Detailed Site Selection Report in Appendix 3).

6.3.3 DESILTING VERSUS EMBANKMENT RAISING

After the topographical survey was carried out and the report (topo drawing) made available for further study and design, the following computations were carried out.

a. Desilting

The cumulative reservoir area under contour lines 21.0 m to 24.5 m was calculated in the course of determining the design reservoir capacity; this was found to be 130,720 m^2 or 13.1 ha (see Table 6.7, Area/Capacity Data for Kishangura Dam Rehabilitation).

TABLE 6.7: KISHANGURA DAM – AREA/CAPACITY DATA

LEVEL	ELEVATION (m)	AREA (m²)	(ha)	CAPACITY (m³)
1	20.75	-		-
2	21.0	3,075	0.31	769
3	21.5	11,396	1.14	4,387
4	22.0	30,179	3.02	14,780
5	22.5	49,858	4.99	34,790
6	23.0	68,278	6.83	64,323
7	23.5	92,048	9.20	104,405
8	24.0	111,367	11.14	155,258
9	24.5	130,702	13.07	215,775
10	25.0	145,206	14.52	284,752

Assuming that one intends to deepen the reservoir by 0.5 m, i.e. desilt, the volume of excavation is 65,360 m³ (130,720 m² x 0.5 m).

Sharing this depth of excavation between clearing/grubbing and desilting, the two different depths are:

i. clearing/grubbing = 0.15 m, giving a volume of 19,608 (0.15 x 130,720 m² or 13.1 ha)
ii. desilting = 0.35 m depth, giving an excavated and carted volume of 45,752 m³ (0.35 m x 130,720 m²).

Using the engineer's rates[15] for Clearing/grubbing and Spillway excavation for Desilting, the cost of increasing the reservoir Capacity by 65,000 m³ = A + B where A = cost of (A) above and B = cost of (B) above.

Hence, A + B = 13.1 ha x Sh. 480,000 + 45,752m³ x Sh. 4,900
= Sh. 6,288,000 + Sh. 224,184,800
= Sh. 230,472,800

This gives a cost per unit volume (shilling per cubic metre) of Sh. 3,526.21 (i.e. 230,472,800/65,360). This cost seems prohibitive, hence the other option is considered – raising the embankment.

In both cases, the same amount of work has to be carried out in the spillway and since this is common to them, it is not a subject of this comparative study.

b. Embankment Raising

Because the embankment level prior to rehabilitation varies from elevation 24.3 m to 24.6 m (i.e. height 3.3–3.6 m), the original elevation of the embankment crest can be approximated to about 25 m.

The design objective is to raise the embankment by 0.5 m, i.e. to 25.5 m, and fix the spillway control section at elevation 24.5 m and compare the quantities and costs with those of desilting. The cross-sections of the

---

[15]Engineer's rate was used for the pre-contract analysis

original and design embankments are shown in figure 6.8.

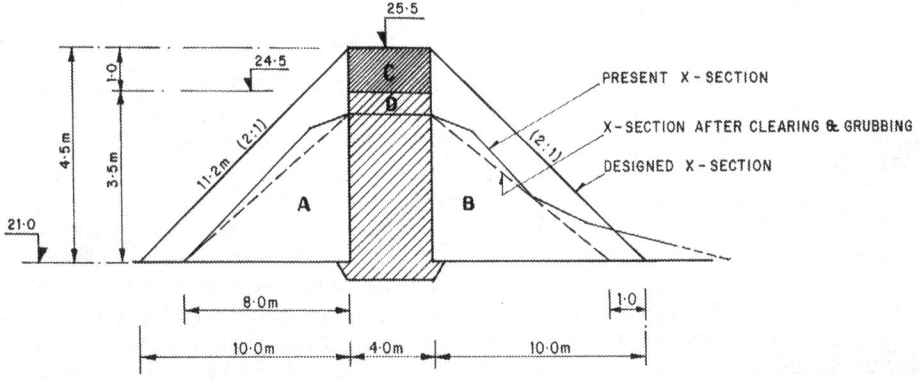

Figure 6.8 Kishangura Dam – Maximum Cross-sections

The volume of earthworks involved in this option is 2000 m³ (See Kishangura Dam – Calculation of Embankment and Spillway Quantities in Appendix 3).

The material required for raising the embankment is homogeneous; it can therefore be rated as for a dam wall, for which a contractor quoted Sh. 6,000 per m³. This gives a total cost of Sh. 12,000,000, and an additional reservoir storage (capacity) of 56,250 m³, giving a cost per m³ of additional storage of Sh. 213.33 (i.e. 12,000,000/56,250, see Table 6.7.)

Note that if grubbing of the reservoir is presumed to be carried out along with embankment raising, this gives an additional storage of 0.5 x 130,720 = 65,360 m³; and if paid for at Sh. 480,000 per ha, as in the case of desilting, the cost will go up by Sh. 6,288,000 for the 13.1 ha surface area of reservoir. However, this is not desirable and therefore should not be done because it would open up the reservoir to erosion and premature siltation and therefore a shorter life span. Whereas grubbing cannot be avoided in the case of desilting.

c. Conclusion
From the foregoing analysis, the unit cost of increasing reservoir storage capacity by raising the embankment is about 8.5% of the contemporary cost by Desilting, or in plain language, Desilting is about fifteen(15) times more costly than the option of increasing the height of the Kishangura Dam embankment.

6.3.4 GENERAL COMMENTS
i. This approach applies to any silted dam, the site potentials of which have not been fully utilised (i.e. where the valley is still rising on both sides

above the present embankment level, and the height of the dam can therefore be increased).

To test the general applicability of the approach, let us apply it to three other silted dams that require rehabilitation: Dyangoma in Mubende, Rwamakara in Sembabule and Kyera in Mbarara Districts respectively (see longitudinal profiles in Appendix 3).

ii. Dyangoma embankment was raised by 2 m from elevation 197 m to 199 m and a new spillway constructed with the Control Section at elevation 198 m, compared with the old spillway at elevation 194.6 m. Rwamakara embankment was raised from elevation 998.6 m to 1,000.5 m i.e. by 1.9 m and the new Spillway Control Section fixed at elevation 999.5 m, about 2 m above the eroded level of the old spillway (997.5).

The additional storage volumes obtained were 136,500 m^3 for Dyangoma and 524,000 m^3 for Rwamakara at additional costs of Sh. 75,279,750 (i.e. 136,000 m^3 x Sh. 551.5/m^3) and Sh.76,980,000 (524,000 m^3 x Sh. 145/m^3) respectively, even when the cost of seepage control was included in both cases.

If these were to be achieved by desilting, the costs would be Sh. 750,750,000 and Sh. 2,882,000,000 respectively, i.e. 10 and 34 times the cost of embankment/spillway raising respectively. These comparative costs are presented in Table 6.8.

As can be seen from the longitudinal profile of Kyera Dam, the potential of that site was fully utilised at inception; in fact, it was a marginally suitable dam site. If and when the reservoir gets silted up, the only option for rehabilitation would be by Desilting.

TABLE 6.8 DESILTING VERSUS EMBANKMENT RAISING[16]

DAM	DESILTING			EMBANKMENT RAISING			REMARKS
	EARTHWORKS VOLUME (M^3)	ADDITIONAL STORAGE (M^3)	COST/M^3 OF ADDITIONAL STORAGE (SH)	EARTHWORKS VOLUME (M^3)	ADDITIONAL STORAGE (M^3)	COST/M^3 OF ADDITIONAL STORAGE (SH)	
Kishangura	65,360	65,360	3,526.21	2,000	56,250	213.33	Desilting is 16 times as costly or 15 times more costly than embankment raising
Dyangoma	136,500	136,500	5,500	Various – embankment and new spillway + seepage control	136,500	551.5	Desilting is 10 times as costly or 9 times more costly than embankment raising
Rwamakara	524,000	524,000	5,500	As above	524,000	145	Desilting is 34 times as costly or 33 times more costly than embankment raising.

[16] The figures obtained for two other dams are included to show that the comparative advantage of embankment raising is not peculiar to Kishangura.

Chapter VII
BID DOCUMENT, BIDDING, CONTRACTING AND CONSTRUCTION

7.1 Calculation of Quantities

7.1.1 INTRODUCTION

After the design has been completed and the design drawings produced, the engineer goes on to determine the quantities of work and materials required to construct his designed structure(s). This exercise leads him to the preparation of:

i   bill of quantities for all items of work involved in the construction of the designed structures, for would-be contractors to use in pricing the scheme and
ii the engineer's estimate, which gives the client an idea of the likely cost of the scheme – discussed in chapter VI.

To represent the three categories of structures, namely new dam construction, rehabilitation of existing dams and construction of valley tanks, the procedure for calculating bills of quantities is presented for three different sites in Uganda.

a. Kenwa New Dam, Mbarara District
b. Kishangura Dam Rehabilitation, Mbarara District
c. Nakakabala Valley Tank Construction, Kiboga District.

7.1.2 EMBANKMENT QUANTITIES

See Figure 7.1 for maximum cross-section and other sketches. Also see Figure 7.2 for the area/capacity curves. Normal water level (NWL) = 50 m, storage = 225,000 m^3. Embankment length = 198 m (Figure 7.3).

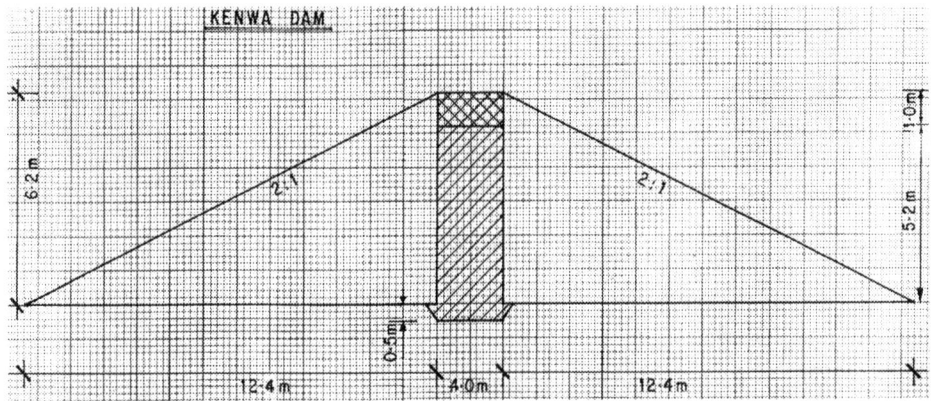

Figure 7.1A: Maximum Cross-Section

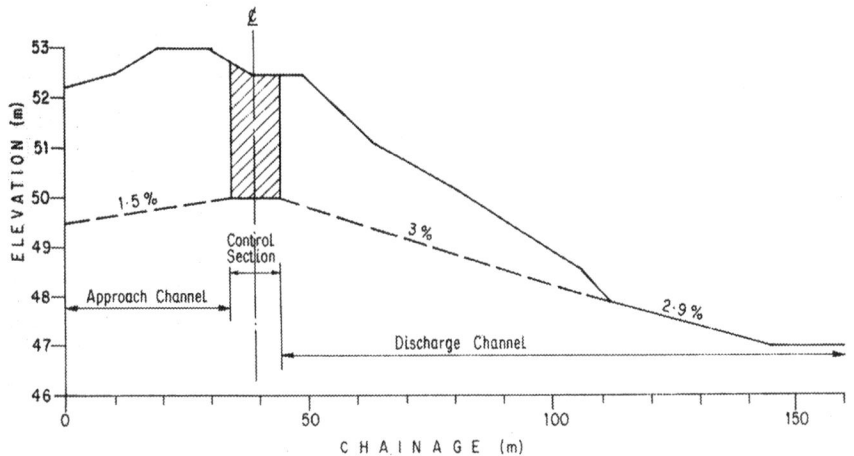

Figure 7.1B: Kenwa Dam, Spillway Longitudinal Profile

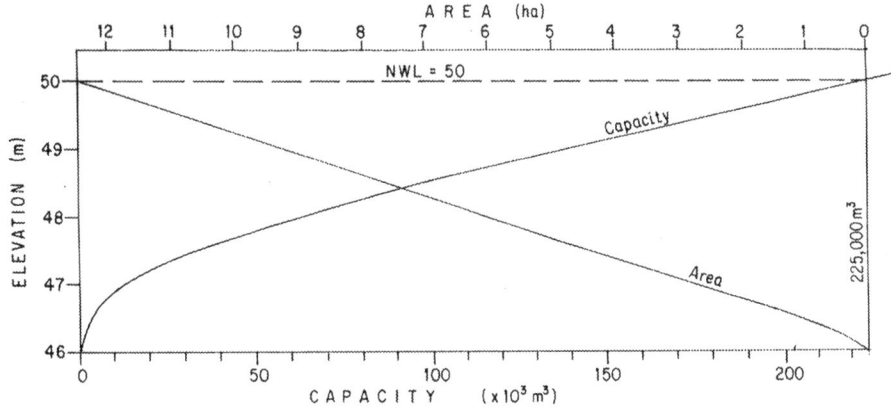

Figure 7.2: Kenwa Dam, Area/Capacity Curves

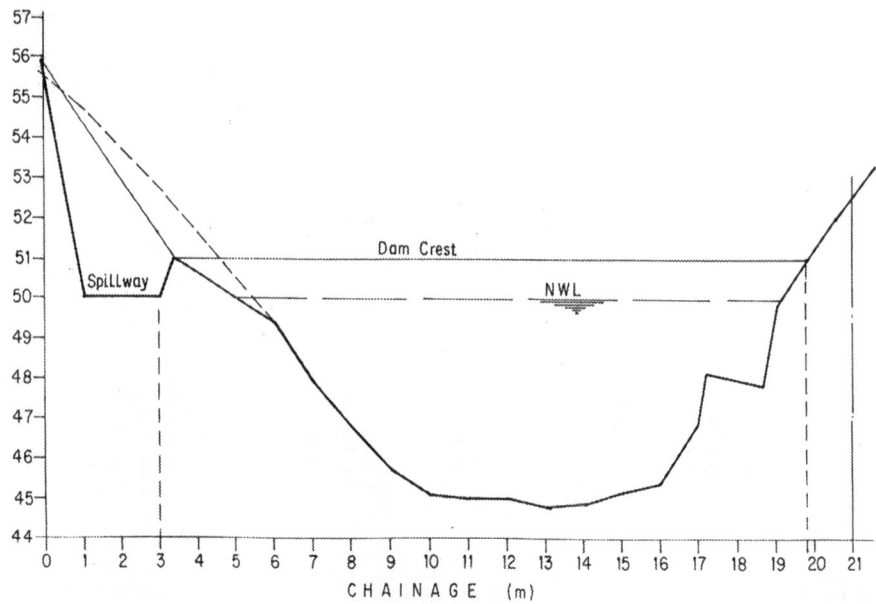

Figure 7.3: Kenwa Dam, Londitudinal Profile of Dam Axis

The volume of the embankment is determined by a step-by-step approach as follows:

a. Calculate the volume of Core foundation by multiplying the depth and average width and length. In the case of Kenwa (fig 7.1), Volume of core foundation = 443 m^3
b. Cut slices of the dam at convenient heights or chainages across the entire profile.
c. Calculate the volume of each slice; the sum of all slices gives the total volume of the embankment. Mathematically, the shell and core volumes are calculated simultaneously in tabular form, using the following equations for each slice:

For the adopted 2:1 slope sided embankment, the volume of each slice is as follows:

$\quad$ Core volume = 4(h-1)d m^3
$\quad$ Embankment volume = h(4+2h)d m^3
$\quad$ Shell volume = embankment volume – core volume
$\quad\quad\quad\quad\quad\quad\quad\quad$ = h(4+ 2h)d – 4(h-1)d m^3
$\quad$ Where $\quad$ h = height of slice
$\quad\quad\quad\quad\quad\quad$ d = depth or thickness of slice

The sum of the volumes of all slices then constitutes the total volume of the embankment in question.

For Kenwa, the computed volumes are as follows:

- Core volume = 2,753 m³
- Shell Volume = 7,614 m³

The slope areas are similarly calculated for each slice and the sum taken.

- Slope Areas = 1,723 m² (upstream)
  = 1,723 m² (downstream)

7.1.3 SPILLWAY QUANTITIES

The selected spillway is trapezoidal, hence either triangulation or conic methods of calculating volumes could be used; the conic method is used here. The volume of spillway excavation is similarly calculated as for embankment. The procedure is as follows:

a. plot the longitudinal profile of the spillway along its centre line
b. determine the elevation (level) of the control section, the slope of the approach channel and the slope(s) of the discharge channel to produce as much as possible, non-erosive flows.
c. cut slices of each section at convenient heights or chainage points across the entire profile
d. the volume of each slice is calculated and the sum of the volumes of all slices gives the total volume of the spillway excavation.

For Kenwa, the volume is 5,172.4 m³. Details of the calculations of embankment and spillway quantities are given in Appendix 4.

7.1.4 LATERITIC MATERIAL CAPPING ON EMBANKMENT CREST

Volume = thickness of capping (h) x embankment top width (w) x embankment length (l) = 0.150 x 4 x 192 = 115.2 m³

7.1.5 REMOVAL OF OLD EMBANKMENT

This site had an old embankment, located upstream of the new dam which has to be removed to increase the storage capacity of the reservoir and for aesthetic reasons.

The embankment is divided into 3 sections of lengths 40, 40 and 60 m, respectively, for the sake of convenience; the average cross-sections for these parts are then used to compute the volume of earthwork, based on the conic method.

Volume a = 1,840 m³, b = 1,808 and c = 2,712
Total = 6,360 m³

7.1.6 OPENING UP ACCESS ROAD
About 1 ha (length x width)

7.1.7 AREA TO CLEAR
i. Between the old embankment and watering troughs:
$$180 \times 210 = 37,800 \text{ m}^2$$
ii. Between the fence line and NWL:
$$15 \times 2,345 = 35,175 \text{ m}^2$$
$$\text{Total (i + ii)} = 110,775 \text{ m}^2$$
$$= 11.1 \text{ ha}$$

7.1.8 AREA TO GRUB
Around the new embankment: $40 \text{ m} \times 230 \text{ m} = 9,200 \text{ m}^2$, say 1 ha.

7.1.9 AREA TO DESILT
This is the reservoir area at NWL less area between the old and the new embankments:
$$123,736.5 - 16,046 = 107,690.5 \text{ m}^2$$
$$\text{Hence, volume to desilt} = 107,680 \times 0.5 \text{ m}^3$$
$$= 53,845.25 \text{ m}^3$$

7.1.10 FENCE PERIMETER
$$= 2,345 \text{ m}$$

7.1.11 AREA TO GRASS
Embankment D/S & U/S slope $= 1,723 \times 2 = 3,446$
Spillway control $= 200$
Filter strip $= 10,900$
Total $= 14,546 \text{ m}^2$

7.1.12 WATER DELIVERY SYSTEM (Figure 6.4A–C)
i. 200 mm dia. PVC class d pipe = 90 m
ii. 50 mm dia. pipe as above = 100 m

7.1.13 OTHERS (see detailed bills)
These quantities are then transferred to a table called the bill of quantities. (Appendix 5)

7.1.14 QUANTITIES FOR DAM REHABILITATION
The major quantities are those related to desilting/embankment and spillway raising as discussed in 6.3. Others are similar to those of dam construction and are calculated as shown in 7.1.3–7.1.13 above. The specific site circumstances will decide which parameters need to be designed and therefore quantities calculated for.

7.1.15 QUANTITIES FOR VALLEY TANKS

The determination of the storage capacities of several tanks was discussed in 5.5 and 6.1. Using Nakakabala Valley as an example, the required calculations are carried as follows.

The volume of water to be stored in the tank, is the same as the volume of excavation to be done to create the storage tank. The volume aimed at for Nakakabala tank is about 19,000m^3, determined in a manner similar to that discussed in 5.5 and 6.1. Because the site conditions allow, the spoil from the excavation is designed to be piled and compacted to impound additional water on top of the tank by creating a spillway with a crest elevation above the ream of the tank.

From equation 3 in 6.1, $C = [A+B+\sqrt{AB}]D/3$ ......3 above
Where
A = Area of tank bottom
B = Area of top of tank and
D = Depth of tank

For Nakakabala Valley tank,

i. Volume of tank C:

$A = 40 \times 60 = 2,400$ m^2
$B = 55 \times 100 = 5,500$ m^2
$D = 5$ m
Storage C $= 19,221$ m^3

ii. Storage above tank ream = Average water depth above tank x average area of the two water surfaces
$= \{[120 \times 130] + [115 \times 125]/2\}1$
$= 15,000$ m^3
TOTAL STORAGE $= 19,221 + 15,000$
$= 34,221$ m^3

Other site-specific parameters, similar to those of dams are similarly determined and the relevant quantities calculated.

7.2 Bills of Quantities

From the calculation of the quantum (quantities) of work and materials to purchase, a format is prepared, called Bill of Quantities, into which all items of work and materials and their quantities as calculated in 7.1 above are entered. Tables 2.1, 2.2 and 2.5 are also examples of Bills of Quantities.

7.3 Engineer's Estimate

In order to give the project owner an idea of the likely cost of the dam and appurtenant works, the engineer prices the bill as best as he can, based on prevailing market prices, especially for materials, and where possible, using the rates of contemporary works in the area; the Ministry of Works is generally helpful in pricing certain items of work. He then sums up the amounts of all work items to arrive at the engineer's estimate. It is customary to add some 5–10 % of the sum arrived at as allowance for contingencies as an insurance against going beyond the estimated cost.

The bill of quantities and engineer's estimate for Kenwa Dam are presented in Table 7. When the last two columns are left blank, the table is called Bill of Quantities, but when they are filled out by the engineer as stated above, the whole table is called Engineer's Estimate.

7.4 Specifications

In designing the dam or valley tank, the engineer has design considerations, which guide his choice of parameters. To ensure that the structure is constructed in accordance with his dream(s), he prepares a set of guidelines to be followed by the contractor or other builders, these are called Specifications. For example,

- the foundation should be excavated to an average depth of 0.5 m, – it should however continue to sound stratum wherever soft formations are encountered,
- the intake sump shall be at elevation 0.5 m above the lowest level in the reservoir, the scour/outlet pipe should be fitted with an aluminium strainer
- at every 3–4 m, a concrete puddle flange (or cutoff collar) should be built around the outlet pipe all through its length under the embankment, etc.

It is these specifications that serve as monitoring/supervision indicators for the owner's agent. Sample specifications and bills of quantities are given in Appendix 5.

7.5 Diversion During Construction

7.5.1 INTRODUCTION

Every dam site has a potential stream diversion need during construction, though the extent of the problem/need varies with the flood potential of the stream. At some sites, it may be costly and time consuming, while at others, it may be avoided altogether by properly scheduling/timing construction to fall essentially within the dry weather. This problem affects both the economy and scheduling of the dam.

A good diversion scheme strikes a compromise between the cost of diversion and the risks involved. It minimises the potential for serious flood damage to the work in progress at minimum expense. The following factors affect the diversion scheme to adopt:

1. Stream flow characteristics
2. Size and frequency of flood to divert
3. Method of diversion and
4. Specification Requirements.

7.5.2 DIVERSION/CARE OF RIVER FOR SMALL EARTH DAMS

For small earth dams in semi-arid areas, care of river/diversion can be minimised or altogether eliminated by timing construction to fall essentially within the dry spell, which lasts for upwards of three months in most places. However, for emergency's sake and for completeness, the following should be taken note of regarding stream/river diversion.

i. For embankment dams, where large areas of foundation and structural excavation are exposed and or where overtopping of the embankment under construction may result in serious damage and/or loss of partially completed works, incidental flooding should be eliminated by diversion.

ii. In selecting the flood to be used for diversion design, the following considerations will provide a guide:
   a. safety of workmen and downstream inhabitants and property
   b. duration of construction to determine the number of flood seasons
   c. the cost of possible damage to completed works and/or those still under construction
   d. the cost of delay to completion, including the cost of keeping the contractor's equipment idle while flood damages are being repaired.

iii. Sequence of construction operation – the diversion programme should be capable of incorporation into the overall construction programme with minimal impact and delay.

iv. Coffer-dams combined with temporary diversion channels or conduits are the commonest diversion method for small earth dams, especially when the foundation work can be timed so that it is executed during the low-water or dry season.
   a. It should be remembered that the flood water held behind the coffer-dam must be evacuated in time to accommodate subsequent storms, hence, coffer-dams should be combined with a diversion channel or conduit.
   b. Coffer-dams are usually constructed of materials available at the site, the commonest ones being earthfill and rockfill coffer-dams, whose design considerations are very similar to those of permanent dams of same type.

v. It is general practice to require the contractor to assume responsibility for stream diversion during construction of dam and appurtenant works, especially for the type of dams covered in this handbook. But this requirement must be clearly stated at an appropriate place in the Specifications or notes preceding bills of quantities. Any provisions incorporated into the design should also be mentioned.

The specifications should not prescribe the capacity of the diversion works and the method to be used. Any hydrographs prepared from available records should be included. The contractor's diversion plan should be subject to the approval of the owner. The diversion of the river should be a lump sum item in the Bills of Quantities.

vi. For all practical purposes, the diversion in respect of dams discussed in this handbook can be handled by the contractor, though his plan should be approved by the owner. See Sample Specifications in Appendix 5.

7.6 Preparation of Bidding (Tender) Document

This is the document which presents to the would-be contractor(s) the work to be done, its approximate quantities as prepared by the designer (bills of quantities), how it is to be done (specifications), a draft contract agreement, which is generally in accordance with the rules and regulations of the Federation of International Engineers (FIDIC) etc. A typical bidding document will be too bulky to include in a manual, but the table of contents of a typical bidding document, based on FIDIC regulations, presented here gives an overview of the contents. An abridged bidding document is given in Appendix 6.

TABLE OF CONTENTS

1. Background Information
2. Invitation for Bids
3. Description of the works
4. Conditions of Tender- General
5. Tender Opening
6. Process to be confidential
7. Classification of Tenders
8. Determination of Responsiveness
9. Correction of Errors
10. Conversion to single currency
11. Evaluation and Comparison of Tenders
12. Form of Tender
13. Appendix to Tender
14. Form of Tender Bond
15. Agreement
16. Form of Performance Bond
17. Schedule of Supplementary Information
18. Bills of Quantities
    - General Directions
    - Specific Conditions
    - Bills of Quantities for Dams

7.7 Bidding, Bid Analysis and Evaluation and Contractor Selection

A draft bidding document is usually submitted to the owner of the project for his vetting. In a number of cases, Funding Agencies introduce modifications to these FIDIC clauses to suit their own rules and procedures, this is one reason why funding agencies generally insist on vetting tender documents before the tender is floated. After vetting and necessary amendments, if any, the document is finalised by the designer and the required number of copies delivered to the owner, usually called the employer. The employer then advertises the work for interested contractors to bid for. After the closing date for the receipt of Bids, the Bids received within the bidding time are opened and subjected to analysis and evaluation as specified in the Bidding Document. A Bid Analysis Format is given in Appendix 7 together with tables showing:

- i   Bid prices as read out on opening
- ii   Report of Preliminary Examination of Bids
- iii   Comparison with Engineer's Estimate
- iv   Summary of Recommendations

This concludes the Pre-contract services.

7.8 Contract Award

7.8.1 CONSTRUCTION CONTRACT

The owner then decides which bidder wins and invites him/them for further negotiation. A contract is eventually entered into with this (these) bidder(s) in accordance with the format included in the bid document.

7.8.2 SUPERVISION CONTRACT

In order to ensure that the construction works are carried out in accordance with design specifications, a supervisor (person or firm) is usually appointed by the project owner. The supervisor not only ensures compliance with specifications, but also ensures that contractors get their dues; he also manages the construction contract and acts as a neutral arbiter between the employer and contractor(s), guided by his own terms of reference (TOR). Excerpts from a supervision contract are given below:

> Services To Be Provided By The Consultant
>
> The objectives of Site Supervision are:
> i. to ensure that the construction and rehabilitation of dams and valley tanks comply with the letter and intent of the contract documents by providing the necessary on-site supervision of the works; and
> ii. to provide a reduced level of supervision/inspection during the defects liability period of completed contracts.

Scope of Services

the duties of the consultant are to assist the employer in the negotiation of the construction contracts, to supervise the works and to approve the materials and contractor's workmanship in consultation with the employer. The supervising consultant shall be designated 'project manager' and shall assume the powers, duties and responsibilities of that position as defined in the contract.

Detailed Terms of Reference (TOR):

These consist essentially of the following:

a. to assist the employer in negotiating the construction contracts and issuance of the letter(s) of acceptance
b. to issue the order to proceed to the contractors
c. to provide detailed site supervision and quality control of the contractor's workmanship in compliance with the provisions of the contract
d. to ensure that the contractor's bonds, guarantees and insurance comply with the requirements of the contract
e. to verify that the contractor's equipment is suitable for the work for which they are proposed and to continue to monitor the performance of such equipment throughout the contract(s)
f. to approve the contractor's on-site superintendence, key personnel and construction programme
g. to approve the land to be occupied by the contractor, his materials and their sources
h. to approve the contractor's setting out of the works
i. to approve the contractors' construction drawings, in accordance with the provisions of the contract
j. to explain or adjust ambiguities or to correct discrepancies in the contract documents
k. to monitor and appraise the progress of the works;
l. to issue variation orders, evaluate such variations (if any), to fix rates for unpriced work items, all with the approval of the employer and to recommend to the Employer, alternative work methods as may be found necessary
m. to inspect and approve the contractor's interim certificates for payment and to certify the completion of the works or parts thereof as provided in the conditions of contract
n. to hold regular and frequent site meetings with contractor's representatives and to keep records thereof

o. to keep the employer apprised of construction activities, the progress of the works and of any problems that may arise from time to time
p. to inspect the works in the course of construction and minimally during the defects liability period and to recommend to the employer on the issuance of the defects liability certificates
q. to advise the employer on all matters relating to claims from the contractor and to make recommendations thereon, including preparing necessary reports and evaluations and participation at arbitration hearings (if these events occur within the consultant's period of engagement, otherwise at additional cost to the employer)
r. to supervise the contractor in all matters concerning safety on site, including the safety of the general public, and the contractor's and consultant's team
s. to liaise with the appropriate statutory authorities as regards the removal or relocation of utilities and services which may be affected by the works
t. provide on-the-job training in all aspects of construction supervision to any employer's staff who may be assigned to the project from time to time
u. The following Reports shall be prepared and submitted at the specified times: monthly progress reports shall be submitted to the employer (in 10 copies), not later than 15 days after the end of the month to which they refer. They would cover at least the following points:
   i. contractor's site staffing during the month
   ii. project activities
   iii. a record of invoices approved to date, and a summarised representation of payments made to the contractor to date; the projected total construction cost for each major item of work and for each contract accounting for any changes affecting the projected total cost
   iv. a record of visitors to the site
   v. construction photographs (at the employer's expense)
   vi. records of site meetings and copies of important project manager/contractor correspondence shall be included in appendices
   vii a record of weather conditions

A final report shall be submitted one week after the issuance of the final taking-over certificate for the last contract package. This report shall detail the major events of the contract including:

- all variations to the original design
- project progress
- problems encountered and their solutions
- details of all claims received including current states
- cash flow
- contractors' staffing history
- consultant's team of himself and the employer's staff history
- photographs etc.

7.8.3 DEFECTS LIABILITY PERIOD SERVICES

Between the issuance of the taking-over certificates and the due date of the issuance of the defects liability certificates, the consultant's counterpart shall inspect the dams as frequently as may be necessary to monitor the condition of the works. The consultant will however join his counterpart for inspection as soon as the dam reservoirs are filled.

Final inspection shall be made by the consultant and his counterpart on the one hand, and a representative of each contractor on the other, two weeks before the due date of the defects liability certificates.

Where repairs are found necessary, they shall be ordered in writing and the work shall be supervised as may be agreed with the employer up to and including resident supervision.

If repairs are ordered as a result of the final inspection, the project manager shall immediately, in writing, extend the defects liability period to cover the completion of the said repairs after which a further inspection shall be arranged.

On the satisfactory completion of any repairs ordered at the final inspection, the project manager shall issue a defects liability certificate and arrange with the employer, the release of the retention monies, bonds, etc., as provided for under the contract(s).

Details of conditions of contract, duration of engagement, remuneration and other contract conditions are usually agreed between the parties, prior to contract signing and commencement of work.

In a construction contract, the interests of both the contractor and the employer are taken into consideration; the following are some of the issues which cater for the employer:

- Performance guarantee – an assurance given by a third party that the contractor can perform; this is usually in form of a bank guarantee or insurance bond to attest to the contractor's ability and preparedness to perform, it is usually 10% of the contract sum.
- Advance payment guarantee – it is similar to a performance guarantee but it relates more to the granting of an advance to the contractor at the beginning of the contract to enable him provide certain aspects of his requirements e.g. mobilisation of personnel and equipment; it is usually in the value of the advance granted; 10% is also common.

- retention – this is the holding back by the employer of an amount of every payment (usually 10%) made to the contractor in lieu of faulty workmanship. Half of this amount is withheld until the end of the construction when the contractor is issued a completion certificate while the remaining half is withheld till the end of the defects liability period (usually one year).

On the other hand, the contractor is assured that prompt payments would be made for work done. In that respect, the following two certificates are used:

1. Interim payment certificate – this is the periodic payment certificate which the contractor raises to show the employer how much work has been carried out and on the basis of which he requests to be paid for that amount of work. When certified by the employer's representative, the employer makes the necessary payment. However, this payment is not final; if in the course of time the employer discovers that the contractor has been overpaid (or underpaid) for any part of the works, adjustments can be made in subsequent certificates, hence all payments prior to the final certificate are tagged 'interim'.

2. Final certificate/payment – this is the certificate issued at the end of the defects liability (maintenance) period after the contractor has been released from all liability in respect of the work; thereafter, a final account is presented by the contractor, vetted by the employer's representative and the contractor gets paid. This marks the end of the contract. Sample interim payment certificate formats and actual interim payment certificates from a project are given in appendix VIII.

7.9 Construction and Post-Construction Services

7.9.1 INTRODUCTION

Construction is closely related to design. Often, alterations in the originally proposed arrangements or dimensions may become necessary after construction operations have been initiated because of unexpected variations in the properties of materials or because of unforeseen development in foundations, errors in earlier surveys and/or omissions discovered later. It is because of these usual changes in the design drawings that 'as-built' drawings are prepared progressively as construction continues, and are submitted to the employer prior to the issuance of the completion certificate. Proper construction involves adequate foundation preparation, placement and compaction of materials in the embankment and construction in accordance with design specifications. It is the responsibility of the employer to ensure that the works are carried out in accordance with the specifications, and it is for this reason that the employer generally employs competent firms or individuals to supervise the construction works. Where a supervisor is employed, part of the duties of that supervisor is to vet the payment certificates prepared by the contractor(s) and recommend payment to the employer.

7.9.2 CONSTRUCTION PROCEDURE

There is no hard and fast rule about this but the following is the usual approach:

1. Site possession – following contract signing, the site is formally handed to the contractor by the employer in the presence of the supervisor.
2. The supervisor issues the order to commence to the contractor as provided in the contract. The contractor then commences as follows.
3. Identification of the bench mark (BM) and dam axis.
4. Mobilisation of initial equipment and personnel to site in accordance with the approved work programme.
5. Clearing the bush around the dam axis and spillway profile.
6. Setting out of dam axis and spillway profile and confirmation of ground levels.
7. The supervisor approves the setting out and/or makes necessary adjustments.
8. Grubbing the dam base and foundation excavation.
9. Identification of borrow areas for construction materials – this could even come earlier.
10. The supervisor certifies the foundation levels or orders necessary amendments.
11. Foundation filling and compaction begins as specified, using approved materials.
12. Spillway excavation may progress simultaneously; for dam safety against floods during construction, it is advisable that the spillway construction progresses ahead of embankment construction to safeguard the embankment in case of a sudden flood during construction.
13. As far as the weather and site conditions permit, construction progresses as planned and/or as amended from time to time.
14. Payment of periodic interim payment certificates to enable the contractor to prosecute the works as planned.
15. Issuance of completion certificate, after the receipt of 'as-built' drawings and guidelines for the operation of the dam, marks the end of the construction period and the beginning of the defects liability period.

7.9.3 DEFECTS LIABILITY (MAINTENANCE) PERIOD (POST-CONSTRUCTION)

Construction contracts usually provide for a period (usually 365 days), during which the contractor has responsibility for maintaining the constructed structure(s), righting wrongs which were not detected during construction. To pay for the works/repairs carried out during this period, recourse is made to the retention amount by the employer. If the contractor fails to carry out his responsibilities during this period, the employer uses that money to carry out the repairs/maintenance by someone else.

During this defects liability period, the works are jointly inspected periodically by the employer and contractor; the contractor is requested to carry out the repairs of any defects found. The employer also operates the scheme to derive the benefits for which it was put in place and similarly, all defects are brought to the attention of the contractor for repairs.

At the end of the period and upon satisfactory repair of all defects brought to the attention of the contractor, a defects liability certificate is issued to the contractor. He is now released from all liability and his bonds returned to him.

The contractor then presents a final account; after the approval of this final account by the employer, all payments due to the contractor are made and the parties part in peace.

TABLE 7   KENWA DAM – BILL OF QUANTITIES/ENGINEER'S ESTIMATE

SITE: KENWA   DISTRICT: MBARARA   COUNTY: KAZO   SUB-COUNTY: BURUNGA

WORKS: DAM CONSTRUCTION

NO	ITEM	UNIT	QTY		U.SH. '000
	BILL NO 1				
	Dam Wall & Reservoir – Earth works				
1.01	Open up bushy part of and improve access road (10 m wide)	ha	1.5	LS	1,500
1.02	Clear bush, i.e. trees and shrubs over dam and reservoir area	ha	7.3	500	3,650
1.03	Grub topsoil over dam area	ha	1.0	300	300
104	Ditto for reservoir to 300 mm depth after dewatering and cart to spoil	ha	11	420	4,620
1.05	Excavate foundation trench for dam core	ha	684	4.5	3,080
1.06	Remove present embankment & cart to spoil or reuse	m³	6,360	4.17	26,521.2
1.07	Borrow, cart, fill & compact good clay for dam core – generally available within the valley	m³	2,458	6.0	14,748
1.08	Ditto for dam wall, using good fill material (from reservoir area and within 4 km of fill area)	m³	7,737	3.9	30,172.3
1.09	Cut spillway and cart to spoil/reuse to have a freeboard of at least 1000 mm from dam crest	m³	4,005	4.0	16,020
1.10	Trim to proper design slopes	m³	3,437	0.75	2,577.75
1.11	Provide, lay & compact lightly a 150mm lateritic/gravelly material capping on dam crest, gently sloping towards the reservoir	m³	115	6.5	746.5
1.12	Topsoiling and grassing Provide and lay 50mm depth of top soil and grass as protective cover against erosion and as sand filter in the following areas:				
	a. Dam upstream slope	}			
	b. Dam downstream slope	} m²	13,934	0.75	10,450.5
	c. Spillway control section	}			
	30 cm wide filter strip upstream of reservoir	}			
	TOTAL BILL NO 1				114,388.25

					U.SH. '000
NO	ITEM	UNIT	QTY	RATE	AMOUNT
	BILL NO 2: PC ITEMS*				
2.01	Supply and build through embankment 20 mm diameter PVC pipe at dead storage level for gravity supply & drainage	m	90	20	1,800
2.02	Supply and fix cast iron gate valves for 2.01 (above): 200 mm diameter, 50 mm diameter	no no	1 1	105 50	105 50
2.03	Supply aluminium strainer for 200 mm pipe and fix in sump with fittings for supply pipe	no	1	285	285
2.04	Supply and fix a PVC reducer tee 200 mm x 50 mm to join drain and supply pipes	no	1	100	100
2.05	Construct necessary manholes with lockable covers for gate valves – 1 & 2	no	2	200	200
	TOTAL BILL NO 2				2,5400

*PC items = Price Cost items

NO	ITEM	UNIT	QTY	U.SH. '000 RATE	U.SH. '000 AMOUNT
	BILL NO 3: PLUMPING, TROUGHS AND SUMP				
3.01	Provide, lay and join 50 mm diameter PVC pipe from end of (2.01) above, including fittings	m	100	7.5	750
3.02	Construct 2 No Watering Troughs				
	- Block work	m²	34}		
	- Concrete base 1:3:6	m³	27}		
	- 3 m hard stand around hard core	m³	56}		
	- *in situ* concrete 1:2:3, 75 mm tapering to 50 mm over hard core		}	12,850	12,850
	- Plastering and rendering		12}		
	- External rough finish	m²	34}		
	- Internal fine finish	m²	30}		
3.03	- Install valve-regulated plumbing fittings for supply to troughs	L.Sum	-	-	300
	Construct sump (1.5 m x 1.5 m x 0.5 m) to enclose strainer for drain/outlet pipe				
	- block work	m²	3.0}		
	- Concrete base 1:3:6	m³	0.5}	325	325
	- Hard core	m³	0.5}		
	- Rendering, external & Internal fine finish	m²	7.0}		
	TOTAL BILL NO 3				14,425

NO	ITEM	UNIT	QTY	U.SH. '000 RATE	U.SH. '000 AMOUNT
	BILL NO 4: FENCING				
4.01	2-strand barbed wire, 150 mm diameter solignum-treated eucalyptus posts at 3 m c/cs and 1 intermediate batten strainer at corners, centre and gates	m	2,345	4.8	11,256
4.02	Provide well-framed gate in fence to allow for inspection and maintenance	no	1	80	80
	TOTAL BILL NO 4				11,336

SUMMARY OF ENGINEER'S ESTIMATE

NO	ITEM	SH.
1.	Earthworks	114,388,250
2.	Drainage and outlet works	2,540,000
3.	Plumbing, troughs and sump	14,425,000
4.	Fencing	11,336,000
	Sub-total	142,689,250
	Contingencies (10%) (to be utilised as directed by the project manager)	14,126,925
	TOTAL	156,958,175

Chapter VIII
OPERATION AND MAINTENANCE, MONITORING AND EVALUATION

8.1 Introduction

Operation and Maintenance (O&M) is a set of actions and activities carried out to actualise the project objectives and to ensure that the benefits are sustained throughout the life of the project.

The primary objective of O&M with respect to dams and related water control structures is to release the project benefits, i.e. water for various purposes.

Rather unfortunately, O&M is one of the most underestimated aspects of water resources development, especially in developing countries. An efficient adoption of O&M is an absolute must if the expected benefits are to accrue on time and to specific target groups, viz.:

- water supply to crops, livestock and humans
- electricity generation
- farmers at tail end of the supply system – canals or pipes -receive their planned share of, say, irrigation water
- properly functioning drainage system to avert waterlogging and salinity etc.

One of the problems identified with old water control facilities is that the project agencies, e.g. river basin development authorities, were not ready for O&M when construction was completed! Furthermore, O&M was always accorded low priority with the result that they were not being funded, hence, maintenance activities/efforts got postponed perpetually in most cases, until major crises occurred. The project efficiency thus declined steadily and the problems to be resolved due to years of neglect of O&M became very complex, and even more expensive and technically complex than if O&M had been accorded the right place on a regular basis.

Neglect of O&M is even worse for drainage schemes – leading to salinity and water-logging, though seldom realised in time because it takes a long time for these problems to surface.

8.2 Operation

For effective and trouble-free operation of storage dams and appurtenant structures, the following steps are relevant:

1. A post-construction O&M team should be set up at the inception of the project. The team should be involved in the project from planning through construction to give its members the opportunity of getting familiar with the design and construction considerations and to become aware of the problems that may require special attention during O&M. The team could be paid for its involvement either directly by the owner agency or through the construction contract.

2. A post-construction inspection should be made soon after completion by the design, construction and O&M personnel to ensure that all items are complete or deficiencies identified for later completion (i.e. completion inspection report). This inspection should identify and discuss, among others, problems, unique operations, general maintenance requirements etc. Procedures should be established for the proper handling of the identified problems and issues. The requirements and procedure for handling initial reservoir filling should be established and agreed upon so that upstream populations to be evacuated and resettled are given adequate warning. Extra procedures and precautions for operation should be established because unpredictable situations could occur. During the first filling, the dam and appurtenant structures should be attended cautiously.

3. Beneficiaries should be encouraged to see the facilities as theirs (irrespective of who pays for it), as a means of improving their levels of income and standard of living in general. Based on this, they should:

   i. Look after and protect the facilities against spoilage, abuse and vandalisation.
   ii. Ensure that any restriction imposed on human and livestock movement in the area, e.g. by fencing, is respected so that the scheme is not subjected to premature degradation.
   iii. Protect all fixtures and mechanical equipment, e.g. valves, pumps and livestock watering facilities.
   iv. Form themselves into water-user groups or strengthen existing ones to formulate and enforce bylaws, among other things to:
      a. ensure that the gate valves are properly opened and closed when necessary
      b. ensure that they make token payments towards repairs and maintenance
      c. repair any damage or malfunction promptly and
      d. organise regular inspection and maintenance of the facilities.

Storage dams should be:
4. Operated to provide as many benefits as is feasible.
5. Operated and releases made to provide optimum benefits, considering the contractual requirements and primary benefits – dams could yield benefits related to irrigation, livestock watering, electric power generation, municipal and industrial needs, recreation, flood control, wildlife and fishery.

   Note that not all dams will provide all these benefits, but an evaluation should be made to identify all potential benefits and how best to derive optimum benefits through dam and reservoir operation.
6. Operational requirements should be based on studies and experiences; these requirements should be documented in the written instruction for dam operation.
7. Multiple use of reservoirs often results in conflicts among beneficiaries, e.g. optimum power production may result in reduced irrigation water supply and fewer recreational benefits. Through proper evaluation of possible conflicts, multiple benefits can be achieved without significant loss to primary beneficiaries, if properly planned and operated.
8. Obtaining accurate and timely hydrological data is critical to proper and safe dam and reservoir operation. Safe operation implies:
   i. reservoir evacuation to pass flood flows and
   ii. maximum storage for given conditions

   This requires a consideration of the entire hydrological cycle on a basin-wide basis. Pertinent information required for efficient operation include: quantity of precipitation, distribution over time and relative uncertainty i.e. forecast errors. Relevant consultations should be made on this aspect.

Precautions During Periods of Potential High Flow
i. be guided by forecast inflows, potential run-off, reservoir elevation and downstream conditions. The dam should be continuously operated during this time. It must be under continuous surveillance.
ii. debris should be removed/cleared from reservoir areas periodically, e.g. annually.
iii. protective vegetation on slopes not otherwise protected should be given adequate attention, i.e. protection against erosion, sloughing of banks, to avoid costly maintenance and safety problems.
iv. obtain and follow expert advice on the suppression of algae growth in reservoirs. No chemicals should be introduced into reservoirs without expert advice.

8.3 Maintenance

While operation ensures that the dam benefits accrue to the beneficiaries as planned, maintenance ensures that these and other derived benefits are provided throughout the useful life of the dam. This is done by keeping all dam components in top form, as close as possible to what they were immediately after construction.

A number of O&M activities dovetail into one another; there is therefore no watertight division between O&M activities. Nevertheless, the following are more related to maintenance.

1. Routine Maintenance Inspection of the dam and appurtenances should be a continuous exercise. All unusual conditions that may adversely affect the operation, maintenance or safety of the dam should be reported promptly, using predetermined written procedures e.g. Reporting Formats to be completed on site.
2. Periodic in-depth inspection of every dam should be made at least every 5 years; the depth and frequency should depend on dam size (the bigger, the more frequent), hazard, complexity and previous problems encountered. These inspections should be carried out by a team of qualified professionals, led by an engineer not directly involved in the O&M. The engineer should be accompanied by Operations personnel familiar with all facets of the O&M. Inspections should be scheduled, if possible, during alternate periods of high and low water to observe conditions unique to each period.

    Special inspection should be carried out when there is reason to believe that significant damage has occurred or has the potential to develop.

    The deficiencies found should be adequately reported and procedures established for correction in a timely manner. The responsibility for correction should be properly and clearly documented. Funding schedules should be considered to ensure adequate and timely funding.
3. Underwater inspection of facilities not normally seen should be scheduled for say, every 6 years. If need be, the reservoir should be dewatered to better evaluate the conditions of such facilities – this is better done at the end of the dry season, just before the rainy season sets in to refill the reservoir. The report of such inspection should be prepared in detail, describing the condition of the facilities and citing identified deficiencies.
4. Written instructions should be availed for O&M personnel to operate the dam. These instructions, furnished by designers and manufacturers, should include the procedures for:
    - routine servicing and maintenance requirements for special operation and maintenance equipment

- the procedures generally referred to as Standing Operating Procedures (SOP), should include Emergency Preparedness Plans, Inundation Maps, the extent and nature of inspections, hydrological and reservoir operations and other pertinent aspects of O&M.

The O&M of the dam should be in accordance with these procedures; significant deviations from these procedures should not be made without the written approval of higher management or engineering personnel.

A copy of the O&M procedures should be accessible to the dam operators during routine operation and during abnormal/emergency conditions at the dam.

5. A logbook should be kept for each dam, to record all significant actions, information and events such as releases, seepage, maintenance, emergencies, etc. The logbook should be kept at the dam or other accessible convenient place for ready reference and use. It should become a permanent record for the dam.

    Dam O&M personnel should be trained before their independent operation of the dam. The degree of training should depend on the conditions and hazards at and below the dam.

6. Periodic inspection of reservoir areas should be made to detect slide areas and monitor their progress. Corrective actions should be taken early in these areas to minimise problems. Posting warning signs should be considered if they pose a safety problem to the public, beneficiaries and/or operators.

7. The capacity of a storage reservoir should not be increased by placing stoplogs or other obstructions in an open crested spillway, without reference to the original plans and without the approval of a qualified engineer. Such devices may effectively reduce the ability of a dam reservoir to safely store and pass the predicted inflow design flood.

8. Debris should be cleared from reservoir areas periodically (e.g. annually) and buried in a safe area – not on the riprap areas.

9. Instructions for operating mechanical equipment should be followed closely to prevent damage to the installations through improper operation. Instructions for the control of spillway gates during flood flows into the reservoir should be followed in detail as in the written operational procedures.

The discussions so far refer to storage dams. The following refer to other types of dams.

10. Diversion dams are usually built to raise the water level in the stream to facilitate diversion. They are usually overflow dams, or have long overflow

sections – to divert water into irrigation canals, water treatment plants or to spreading grounds for recharging groundwater. They should be operated, maintained and inspected as for earthfill or concrete dams.

11. Flood detention reservoirs serve to reduce flood peaks by the temporary storage of part of the flow that exceeds the capacity of the spillway or outlet works of the dam. Note that all reservoirs or pools produce some detention effect.

    Common Features of Detention Reservoirs are:

    i. Outlets that automatically control the rate of release within safe limits
    ii. Overflow spillway to protect the dam – even at the expense of possible damage below the structure in the event of floods far in excess of the design flood.
    iii. Outlet works of such dams should be kept free of soil deposits that might affect their proper functioning.

12. Emergency Preparedness Plan (EPP)
    An emergency preparedness plan should be developed for all dams and conveyance facilities, whose failure would endanger human life or cause substantial property damage. The EPP should include all pertinent instructions for a dam operator to follow during an emergency.

    Each EPP should be discussed with the local community leaders or the people directly involved/responsible for the well-being of the citizenry for their comments.

    All relevant authorities – local, state and federal – should be aware of the potential danger a dam failure would present and should be assisted in developing an early-warning system/procedure, especially for the endangered downstream population; the system should include the necessary inundation maps or descriptions delineating flooded areas.

13. To ensure that a dam is properly operated in an efficient and correct manner, the owner is obliged to train each dam operator. The object of the training is to acquaint the operator with the full range of operations required. The training should be in two aspects: general training in the classroom, and on site at the operator's workstation.

    The training should be specific as to the operator's responsibilities. It should provide awareness and working familiarity with all operation documents; it should emphasise the importance of accurate and complete record keeping – entries in forms, concise explanations in diary format, use of tape recorders and taking quality photographs of events and conditions.

    It should provide enough information for the operator to make knowledgeable, correct and prompt decisions concerning the protection of facilities and downstream life and property. The best source of information for all these is the EPP.

    Note that sound judgement is an intangible quality, which can be greatly enhanced through complete familiarity with capabilities and limitations of the physical facilities under the operator's care.

All dam operators should take refresher courses every 3 years. If possible, a prospective operator should be trained before assuming full responsibility – as in an assistant position.

Consolidated Maintenance Activities
1. Maintenance of earthfill dams consists of:
    i. removing debris from the upstream slope
    ii. replacing disintegrated rip-rap or grass
    iii. removing deep-rooted shrubs from both slopes and controlling the occupation of rodents, which could bore holes in the dam
    iv. repairing eroded areas
    v. proper grading of access roads
    vi. maintaining monitoring devices (see M&E) within the embankment and adjacent areas
    vii. controlling vandalism.
2. Maintenance of structures and mechanical equipment
    Two vital features to the performance of most dams are the spillway and outlet works – in one form or another, most dams are provided with these. All components of these features should be included in the inspection of each dam.
3. Outlet works
    Regular inspection should be made of:
    i. Metal and concrete surfaces for abnormal conditions
    ii. Deterioration of protective coatings on metal, which will reduce the effective life of pipe and equipment substantially.
    iii. Small irregularities on the surfaces of passages may contribute to the incidence of cavitation, which can lead to rapid deterioration of metal and concrete.
    iv. Large leakage of groundwater into access shafts, tunnels, gate chambers and control houses can be detrimental to equipment and metal work and can be a safety hazard to personnel.
    v. Cracking of concrete in tunnel linings, shafts and gate chambers should be monitored and any differential movement between adjacent structures should be noted.
    vi. Special inspection of submerged structures should be made, by dewatering or when operation conditions permit. Stilling basins have been the feature requiring the most regular monitoring and major maintenance. Rock and debris should be removed from them regularly.

4. Spillway
    i. Inlet and outlet channels (or approach and discharge channels) should be free of trees and debris that could impede flow. Restrictions in these channels reduce the capacity of the spillway.
    ii. Differential movement of spillway walls, crest and stilling basin should be observed.
    iii. Erosion of slopes within the spillway should be controlled.
    iv. Proper drainage should be ensured.
    v. Stilling Basins should be inspected as for outlet structures.

Mechanical Equipment

i. Some outlet works and spillway depend on the ability of some mechanical equipment to perform, e.g. gated spillways, outlets controlled by radial gates etc. The inspection of the structure should include all aspects of the attendant mechanical equipment – gates, valves, pumps, controls and auxiliary equipment – should be operated and observed during inspections if possible.
ii. Mechanical equipment should be lubricated and serviced according to manufacturer's instructions.

For small earthfill dams in Uganda, based on the author's experience, the guidelines presented in Table 8.1 were prepared for the beneficiaries, who were expected to maintain the facilities, once handed over to them.

TABLE 8.1 GUIDELINES FOR MAINTAINING DAMS AND VALLEY TANKS PREPARED FOR USER COMMUNITIES IN UGANDA

STRUCTURE/ FACILITY	MAINTENANCE ACTIVITIES
1. Dam Wall (Embankment)	- Keep grass growing on the upstream & downstream slopes to safeguard them from erosion; any sign of erosion should be attended to and grassed promptly. - Trees and shrubs should not be allowed to grow on the embankment as their deep roots could create holes in the dam and lead to leakage (piping).
2. Reservoir	- Avoid polluting the reservoir water with human and livestock wastes, alcohol residues (as from distilling local gin at the dam site); no spraying of animals should be allowed upstream close to the dam; the use of fertilisers and insecticides should not be allowed in gardens upstream of the dam. - Maintain reservoirs clear of weeds/vegetation to avoid sedimentation and eventual siltation; seek the advice of experts before using chemicals to destroy aquatic plants. - Under no circumstances should mud troughs be made within the reservoir. - Report to health authorities the presence of disease vectors like snails, which transmit bilharzia to humans.
3. Spillway	- Repair eroded areas of the spillway promptly to avoid further damage. - Ensure that the entire spillway is kept clear of any form of obstruction as this may endanger the dam in case of heavy storm and high run-off. - Avoid grazing within the spillway because cattle hooves could break the soil and encourage erosion. - Trees and shrubs should not be allowed to grow within the spillway as they may damage and block the spillway.
4. Water Delivery System	1. Sump – Remove any silt that might have been deposited in the sump at the end of every dry season so that the strainer in the sump is not blocked by silt. This may require the use of boats or divers in deep reservoirs which do not dry out at the end of the dry season. 2. Gate valve – Ensure that this is carefully opened and securely closed by a designated and trained attendant to ensure that the valve functions properly and does not accidentally empty the reservoir. If the spillway starts spilling and it is still raining, open the valve to help safeguard the dam against overtopping, but remember to close it as soon as possible to avoid draining the reservoir by mistake.

STRUCTURE/ FACILITY	MAINTENANCE ACTIVITIES
4. Water Delivery System (cont'd)	3. Watering troughs – The area around troughs should be kept clean and dry; avoid water pollution and foot rot disease in animals; for this reason, drains should not be allowed to get blocked. - Animals should not be sprayed in trough areas to avoid water pollution. - Repair any damage to troughs promptly to ensure long life. - Taps, Valves and Pipes – Ensure that water taps, pipes and valves are not damaged by animals, children and others. Broken ones should be replaced or repaired as the case may be to avoid wastage.
5. Fence	- Ensure that the fence, gate and barbed wire are not broken to allow animals and unauthorised humans into the dam area, reservoir etc. - Strengthen the barbed wire fence with live hedge, using drought-resistant species. - Animals should not be allowed to graze within the dam area or water directly from the reservoir to pollute the water and damage other structures.
6. Catchment Area	- Check the development of any gullies in the catchment area, close to the dam and repair promptly to discourage erosion and the consequent silting up of the reservoir. - Do not cultivate land upstream, close to the dam/reservoir so that erosion may not wash silt and pollutants into the reservoir – this will reduce the useful life of the reservoir and pollute the water, making it unsafe for humans and livestock consumption. Any unavoidable cultivation must be contour farming and in consultation with your extension officer.
7. Funds	- Raise funds/revenue for purposes of repair and maintenance of your dam; open a bank account for this purpose.
8. Bylaws	- Formulate bylaws for proper management and sustainability of your dam and educate all users accordingly.

8.4 Monitoring and Evaluation

8.4.1 MONITORING

This is a continuous or periodic surveillance over the implementation of an activity – and its various components – to ensure that input deliveries, work schedules, targeted outputs and other required actions are proceeding according to the plan.

The objective of monitoring is to achieve efficient and effective project performance. It is therefore an integral part of management information system (MIS) and an internal activity.

In the case of dams and reservoirs, the periodic examinations and evaluation help to detect/discover conditions that can disrupt operations or threaten dam safety, early enough to be corrected.

8.4.2 EVALUATION

This is a process which attempts to determine as systematically and objectively as possible, the relevance, effectiveness and impact of activities in the light of their objectives.

It is a learning and action-oriented management tool and an organization process for improving activities still in progress and future planning, programming and decision-making.

In addition to evaluating the achievement of dam objectives, safety evaluation of dams is conducted to assess the condition of the dam in relation to its structural and functional integrity.

Evaluation also identifies existing/potential safety deficiencies, which can be confirmed or dismissed by analysing the collected data.

There are four levels to M&E in water development projects; these are:

1. planning, design and construction of physical facilities
2. operation and maintenance for irrigation, drainage and hydro-power
3. agricultural production and
4. achievement of socio-economic objectives.

Level 1, planning, design and construction, is the easiest to handle; it is also a discrete phase, completed once the construction of physical facilities is finished. The other three require continual monitoring and evaluation during the project life to ensure that the system operates at the desired efficiency and that the project objectives are met continuously.

1. Planning, Design and Construction of Physical Facilities
   M&E has always been a standard practice at this level. Engineers and surveyors have always carried out M&E on:

   i. planning and design progress to ensure timeliness and adherence to allocated funds.

ii. use of equipment and construction materials to ensure equitable usage.
  iii. construction, ensuring that it complies with approved specifications and schedule.
  iv. project costs, ensuring that budget estimates are not exceeded.

Additional areas which require M&E are:

  v. Employment creation – for all categories, including the unskilled poor. Opportunities for this should be sought and maximised to stimulate local business and development. It is necessary to monitor the wages paid to men and women to ensure that women are not underpaid on bias. Child labour should be avoided.
  vi. Equipment used.
  vii. Materials.
  viii. Beneficiary participation in planning, especially on issues like canal/pipeline alignment.
  x. Participation of local authorities on issues which affect long-term sustainability of any project. The project should not be seen as an imposition on the local government by the central government.
  xi. Provision of other infrastructures to enhance project outcome, e.g. all-season roads to project sites for easy movement of inputs and evacuation of produce.

2. Operation and Maintenance of Water Control Facilities (this is covered earlier in this chapter).

3. Agricultural Production
   The fundamental objective if irrigation is to provide efficient water control for increased agricultural income. In addition to efficient water application, there are other simultaneously required inputs like seeds, fertilisers, pesticides, machinery, extension services and marketing facilities; all must be availed on timely basis.

   For M&E at this level, all the factors mentioned, except irrigation water (considered at the previous level), must be considered.

   Information must be considered at critical times for each cropping season, which can be used to provide better coordination between the provision of various inputs and services. At the end of the season, an overall performance review of the season is made, which will be useful in preparing an integrated and improved plan for the subsequent cropping season. M&E at this level generally requires maximum efficiency, compared with the other levels.

4. Achievement of Socio-economic Objectives
   The fundamental objective of irrigation is increased agricultural production – increased food availability and increased income of both farmers and non-farmers. These together go a long way to achieve the socio-economic objectives of the project.

The impact of the project on the proposed beneficiaries should therefore be measured on a continual basis, so that decision-makers are aware of the developments in order that appropriate policies may be formulated and implemented on time to reverse undesirable trends. Both the intended and anticipated impacts must be monitored. But socio-economic variables need not be monitored as frequently as those of O&M; 2–5 years frequency will suffice.

The overall aims of maintenance and evaluation are:
1. To ensure that timely corrective actions can be taken for maximising project inputs and so achieve the project objectives.
2. The goal achievements can be determined.
3. Lessons can be learnt for more effective project design and management.
4. Project assumptions can be verified.
5. Overall project impacts can be analysed.

The Principal Requirement of Monitoring and Evaluation
1. Timeliness – information must reach decision-makers at the right time so that the necessary follow-up decision (i.e. conversion of M&E information to action) can be effective. In addition to good M&E information, its quality and proper channelling are equally important. Untimely M&E information could lead to one of the following:
   a. wrong decision
   b. decision taken may not be optimal in terms of the agreed objectives.
   c. No decision may be taken when one is essential
   d. Decision taken may result in irreversible damages or
   e. Decision taken may unnecessarily increase the cost of the project and/or time required for completion.
2. Cost-effectiveness. Information collection involves time, money and expertise. The general rule to remember is that the value of the information collected must be higher than the cost of obtaining it.
3. Maximum Coverage.
4. Minimum Measurement Error.
5. Bias-free. For water development projects, the use of interdisciplinary individuals who, though specialised in one or more disciplines, are flexible, observant, sensitive and are capable of intermixing and questioning inventively is essential. However, a multidisciplinary team is more realistic as it is difficult to find such an interdisciplinary individual; to achieve this, every effort should be made to assemble a formidable team of experienced, committed and disciplined professionals of high integrity and tenacity of purpose.

Safety Evaluation of Existing Dams (SEED).
SEED evaluations are carried out to determine the condition of the dam in relation to its structural and operational integrity. There are two phases: examination and analysis.

i. Examination identifies existing and potential dam safety deficiencies as determined in a review of the design, construction, operation and performance data and from an on-site visual examination of the dam.
ii. The analysis phase evaluates the recommendations made, based on examination, to determine its relevance/significance to dam safety; it goes further to identify necessary actions to resolve the deficiencies, among others.

A comprehensive report of each phase should be prepared however, which should contain: findings, conclusions and recommendations. The reports should be objective, comprehensive, unbiased, straightforward and timely.

The dam examination team generally consists of experienced engineers and geologists, equipped to critically assess the performance of dams during the past and anticipated events.

The findings and recommendations of SEED should be taken seriously by the owners, who should also ensure that necessary corrective actions are taken in a timely manner.

Chapter IX
WATER RESOURCES DEVELOPMENT AND ENVIRONMENT MANAGEMENT

9.1 Introduction

As seen in the hydrological cycle, water exists within the environment constituted by the atmosphere and land as principal components. Modern water resources development must take environmental impact assessment into account in its planning, design and implementation to ensure that the environment remains healthy for the continued life and activities of man, animals and plants.

In more details, planning for environment management is important for the following reasons:

i. to conserve and use the environment and natural resources for the benefit of present and future generations
ii. to maintain ecological processes essential for the functioning of the biosphere, aimed at optimising the sustainable yield of the natural resource – water in this case
iii. to establish adequate environmental protection standards and monitor changes in them
iv. to make prior environmental assessment of proposed activities which may significantly affect the environment or use of the resources therein
v. to give early warning to all persons likely to be significantly affected and
vi. to ensure that conservation is an integral part of the planning and implementation of such developmental activities.

The drainage basin, which is a natural terrestrial/hydrological unit constitutes a convenient unit for the desirable integrated land and water resources development, much better than political and geomorphological units.

9.2 Environmental Impact Study/Assessment (EIA)

The objective of EIA is to ensure a good balance in the development of water resources and the protection and enhancement of environment quality.

Projects need to be multi-objective and have a serious focus on the potential environmental consequences of development. This must be from initial planning and design through construction and operation. It requires an interdisciplinary approach, involving economics, engineering, design, biology, recreation, hydrology and sociology – all focusing on the natural and physical resources and how best to harness them to obtain maximum benefits, while maintaining and improving the environment. Mitigating actions after the project is completed should be avoided – it is medicine after death!

In conducting an EIA, the following points should be noted:

1. Enhancing existing resources and complete avoidance of adverse environmental effects are not always possible.
2. Benefit to one resource may result in loss of another resource, e.g. impounding a stream, which creates a dependable water supply, automatically results in loss of terrestrial resources with area permanently inundated. It is therefore the job of planners (team) to develop plans that minimise negative impacts and maximise positive ones. In most cases, negative impacts can be significantly reduced through careful design, construction and O&M of project features.

9.3 Suggested Steps to Enhance the Attainment of Low Negative Impacts

1. Consult relevant literature and experts in environmental issues to find creative solutions to lessen adverse impacts.
2. The issues discussed here are those commonly encountered in all water resources development projects. But each study should identify the issues relevant to it and thus define its own scope.
3. Consult appropriate agencies, e.g. Federal Environmental Protection Agency (FEPA) in Nigeria and National Environment Management Agency (NEMA) in Uganda, for compliance procedures – there are specific guidelines for Fish and Wildlife Service, Forestry Service and Environmental Protection Agency etc. in the USA, plus water resources agencies.
4. Consult the local population – the oldest inhabitant's experience is always very helpful.
5. Categories of resources to consider include: air, water quality, prime and unique farmlands, wild and scenic rivers, geologic formations etc.

    a. Ecological and Environmental Conditions for Fish: Fish types, population, effect of stagnancy compared with flowing river/stream, water's physical and chemical properties, fish passage between upstream and downstream areas, e.g. the use of fish ladder (a series of stepped pools separated by weirs etc.); retaining some trees and shrubs in the reservoir area as a cover and feeding areas for fish etc.

    b. Wildlife. Impacts could be due to loss and modification of their habitat

and disruption of movement patterns, due to direct and indirect actions like inundation.

c. Water Quality. The quality of the water to be impounded should be considered in relation to its intended use. If the quality is found to be poor for its intended use(s) and anticipated benefits, the reservoir should be considered a failure. This assessment should be an integral part of the planning and design process to avoid water quality-related failures. For drinking purposes, for aquatic life, irrigation, livestock watering etc., each has standards to guide planners and designers.

9.4.1 Effects of Design and Operational Criteria
(BOTH IN THE RESERVOIR AND DOWNSTREAM)

Temperature regime, total dissolved salts (TDS), location of outlets etc. Bottom withdrawals result in cooler water downstream and warmer water in reservoir and vice versa for withdrawal from the epilimnion (from near the surface).

9.4.1 DESIGN CONSIDERATIONS

The primary factor in controlling water quality is site selection, hence:
- obvious sources of pollution should be avoided.
- use outlet to control quality between reservoir and downstream, i.e. either drawing from deep depths or from epilimnion areas.

9.4.2 RECREATION

Water attracts people as a recreation medium i.e. for swimming, fishing and boating, picnicking, camping and sightseeing. This is seldom a major purpose of small dam projects, but it can make an important contribution to overall benefits.

9.4.3 RESERVOIR OPERATION

This has the most significant effect on the location of recreational facilities. For irrigation and flood control dams/reservoirs, the extent, duration of water level fluctuations determine the location of recreation features, less so for hydro-power reservoirs.

9.5 Watershed or Drainage Basin

The drainage basin (or watershed or river basin) constitutes a natural territorial unit of integrated land and water management. It provides a defined geographical/hydrological manageable ecosystem within which eco-processes can be monitored/controlled. The drainage basin is an integrated system that transforms precipitation, solar radiation, other environmental variables, labour and capital in wood products, wildlife, recreation and aesthetic satisfactions and water.

The forest management subsystem, the grazing subsystem, the recreational use and development sub-system and the water management subsystem interact to produce vegetation, animals and soil conditions that govern the yield and quality of its products and services.

As population pressure, agricultural production, urbanisation and industrialisation develop in a river basin, so do the water demands and the risks for land and water degradation.

The most commonly listed/mentioned environmental hazards relate essentially to land degradation, water, forest and soil resources. They all arise from disequilibria caused by human intervention in the original equilibria, i.e. man's inability to fit properly into the natural ecosystem conditions prevalent prior to his intervention.

Basin-wide development strategy calls for the integration of water, land use and environment management for optimal and sustained realisation of desired benefits.

Land–water interdependence implies:

- land conservation is practised to reduce erosion
- water conservation is practised to support land use, especially agriculture.

Integrated development therefore implies flood mitigation, promotion of multi-purpose benefits of water resources development and maintenance of productivity of nature and man.

9.6 Environmental Integration of Land and Water in a River Basin (Watershed)

Land and water are two basic resources in the life-support system on which man depends. Due to increasing pressure on them as a result of population growth and intensification of human activities, there is a growing concern about the environmental hazards associated with development, if allowed to continue unchecked – the ultimate danger being depleting the resource base (i.e. land and water), and deprivation of the long-term carrying capacity of the environment.

Rural development is a predominant exponent for water and land utilisation in terms of its sectoral demand and real extension. But the intensification of land use or changes thereof is quite often a result of urban pressure and overall population increase.

Management is perceived globally as a principle which:

i. assures efficient resource utilisation and
ii. encompasses the necessary conservation of their productivity in a long-term perspective.

Water management has become a matter for primary concern, but due to the complementarity of land and water, management of land resources should be a decisive consideration in water conservation and management.

Note that:

1. With increasing socio-economic development, it is necessary to pay due attention to environmentally sound planning and evaluation procedures.
2. By applying an ecosystem view on resource utilisation, the potentials and limitations of the system can be rationally assessed with regard to sustainable utilisation. Projects should therefore be conceptualised from a strict environmental viewpoint – both new projects and the improvement of old ones – to ensure a long-term realisation of the benefits and a cost-effective financial investment.

9.7 Causes of Resource Degradation

1. The major manifestation of land degradation in most developing countries is soil erosion, caused by:
   - extensive clear felling for agricultural and other purposes
   - overgrazing and
   - improvident use of marginal lands for agriculture.

It is important to note that:

- Land management problems are caused essentially by erosion, waterlogging and salinisation – all caused by water.
- The most serious threat to land stems from deforestation, denudation and soil erosion.
- Erosion implies a progressive loss of productive power through progressive loss of topsoil, which in turn leads to siltation of reservoirs, which represents a valuable irrigation potential, flood control and hydro-power generation.
- Displaced soil raises river beds and thereby reduces their carrying capacity and thus leads to floods.
- Excessive run-off along denuded slopes reduces its percolation into the soil/subsoil strata and results in loss to the sea, often after causing floods – a resource that would have been available for year-round use as groundwater! Floods and droughts are therefore two sides of the same coin of poor land management and conservation of the soil, which automatically results in the conservation of water also.

Degradation sets in when the environment can no longer re-establish equilibrium after man's intervention. Degradation implies that the rate of disruption exceeds the outmatched rate of recovery of the ecosystem. The major aim of environmental planning and management is either to prevent disharmony between man's demands and the constraints of the environment or minimise the adverse effects of man's demands on the environment.

Under better land management conditions, the fresh water lost to the sea as floods, could have been retained as groundwater and put to beneficial use in areas remote from rivers and watercourses.

2. The second cause of Land Degradation is the uncontrolled water logging and salinisation of the soil in canal-irrigated areas due to seepage from unlined distribution systems in easily permeable soils, as a result of not providing adequate drainage. Uncontrolled and often excessive application of water to lands, not properly drained. Waterlogging and salinisation represent a very serious threat to land resources because they affect good agricultural soils into which large investments have been made – expensive irrigation projects, which have potentials for sustained production under conditions of good land and water management.

Unfortunately, most developing countries are not yet aware of the gravity of the threat which the continued mismanagement of their most basic resources (land and water) poses to their survival. For their future, it is important that the situation be brought under control before reaching the point of no return!

3. Water Quality Deterioration
This is a global problem – pollution of rivers by industrial by-products, thus having a negative impact on fishing and other beneficial uses of river as a resource.

New generations of quality problems result from modern agriculture supported by high-level use of fertilisers – often indiscriminate use, caused by blanket application without soil test to determine soil carrying capacity, herbicides and insecticides.

Airborne pollutants are also causing serious water quality problems, including acidification problems in temperate areas with poor soil buffering problems.

9.8 Criteria for Land–Water Integration in Environment Management

1. Water and land are closely interrelated in the natural environment. Most of the water in a river has earlier passed over land; most human land uses are water dependent, but influence water flow, its seasonality and quality. This makes the water divide fundamental for the integration of land and water conservation and the drainage basin an appropriate unit for such integration.
2. Both water and land interventions by man produce changes in the natural

environment; therefore, environmental protection as an isolated entity is not meaningful. Environment management should therefore be seen as inseparably connected with land and water management – two sides of the same coin again!

3. Unavoidable environmental changes due to land and water development should be seen as the price to be paid (opportunity cost) for improved quality of life. The degree of harmfulness of these changes depends on how the productivity of the resources can be sustained. A realistic goal to aim at is therefore an environmentally sound management of land and water, i.e. harmonisation in the development of the environment and socio-economic interests in a basin.

4. The role of environment management should be to maintain the productivity of the primary resources and their linkages, i.e. coordination of land and water conservation and management should aim at optimising the net production of the land. Land use patterns must be based on actual levels of fertility and the possibilities of compensation for the various losses incurred through intensified land use.

5. Developing a basin to maximise water availability and conservation and management of water for maximising socio-economic benefits in a basin are two different issues.

Complementarity of Land and Water Conservation

An integrated management approach is advocated for the most valuable life-support systems – land and water. This consists of conservation and optimum utilisation of both, remembering that water is replenishable annually, whereas land is not. It is non-renewable and irreplaceable; it therefore deserves greater care and attention than water. But because they are intricately interrelated, looking after the land implies looking after water concurrently.

Integrated Land–Water Management – Means and Approaches

1. Conservation and optimal management of land and water is one of the greatest challenges facing mankind; this is more evident in developing countries which face acute problems of poverty, increasing population pressures and land degradation. A sound conservation and management policy offers the only hope.

2. Land use and soil quality may have influence on water flow and quality and vice versa, hence good resource management becomes imperative.

3. Land and water management are inextricably linked, their conservation and development therefore need to be in an integrated manner at all levels, from field to government. Therefore, adopt strategies which seek to accommodate human activities within the purview of natural processes in a way to guarantee their continued use and preservation/conservation by man at minimum cost.

9.9 Crucial Action Components to Achieve Integrated Land–Water Conservation and Management

1. Conservation of irreplaceable soil resources, affected or threatened, through appropriate means like afforestation, restoration of pastures, anti-erosion works, drainage etc.
2. Conservation of water through appropriate soil conservation and land management practices through the construction or restoration of innumerable small local storages in mini-watersheds before impounding genuinely surplus and silt-free waters in large reservoirs.
3. Command area development works (including drainage) in canal-irrigated areas.
4. Use of lands according to their capability, i.e. forestry, pastures or cropping, for the needs of society and in a manner which ensures that their carrying capacities are not exceeded, while their productivity is built up over time.

    The use of marginal lands in intensive manners which result in low and diminishing returns should be stopped/avoided. Similarly, soils should not be overburdened with salts in form of fertilisers in a bid to forcibly improve their productivity.
5. Active involvement and participation of local populations – this is the only way to succeed in the measures mentioned above.

The following actions are a sine qua non for success in the action components listed above:

- Education and training – formal and informal down to village level, involving farmers and non-farmers alike.
- Pay attention to the relative economic productivity of water in different uses as a guide to optimal allocation of water.
- Evaluation and monitoring – for both existing projects in relation to their original scope and for establishing a system for the permanent control of land degradation.
- Technical components –
    - avoid large-scale development projects – they are one of the obstacles to integrated land–water management
    - in relation to urban growth, include measures to improve water supply, manage water demand – water saving, re-use, recycling etc. – and manage water quality.

9.10 Options for Improvement in Land Productivity

1. It makes better economic sense to quickly put existing irrigation potential to optimal use and save precious land from damage by waterlogging and salinisation than to create additional capacity which could end up being under-utilised as well as pose grave hazard to the soil.
2. Irrigation with Groundwater intrinsically lends itself to development by individual farmers at little expense and in record time; but the attraction also creates the danger of over exploitation, leading to abnormal lowering of the groundwater table and juxtaposition with saline aquifers in certain areas.

A systematic investigation into the nature and recharge characteristics of the intended aquifer and the imposition of necessary controls over withdrawals should be undertaken whenever necessary.

The replenishment of groundwater must be maximised by taking comprehensive afforestation and soil/water conservation measures in the watersheds.

Advantages of Groundwater Utilisation (over surface water as sources of irrigation water)

1. No expenditure needs be incurred on either the storage or the conveyance.
2. No land needs be acquired for either the reservoir or canal systems.
3. No seepage losses – often as much as 50% of the water released from reservoirs gets lost in transit due to seepage!
4. Groundwater is not susceptible to evaporation losses – either during storage or transmission.
5. Surface water systems have long gestation periods – upwards of two years in some cases – in the course of dam construction, canals etc., whereas tube wells can be installed and commissioned within weeks, if not days.
6. Land shaping for groundwater irrigation can be done by the farmer himself; in canal irrigation areas, such works must be done on the basis of entire outlet commands.
7. Problems of water distribution and drainage – waterlogging and salinisation are almost non-existent in groundwater-based irrigation areas because there is no seepage from canals, no interference by canals with natural drainage of the area and because the farmer can easily plan his own distribution system and is careful not to apply excess water to the land.

For these reasons, the possibility of using groundwater should first be explored before resorting to surface water.

9.11 Sedimentation in Reservoirs

9.11.1 THE UTILITY OF A RESERVOIR

This diminishes as its storage capacity is reduced – usually by sedimentation. Soil losses due to erosion, especially in semi-arid/savannah areas are very high; for example, in Bawku District of the Upper East region of Ghana, losses are in the order of 1.56 ton/ha for run-off arising from a 10 mm rain for a tillage situation of 50% ridging across the slope for early millet and guinea corn rotated with groundnuts (Halm, A.T. & Asiamah). With the average annual rainfall of 1,100 mm and an estimated total surface run-off of 12.5%, the estimated annual soil loss (erosion) would be as follows for two water-related situations:

- dams with irrigation: for overall mean size catchment area of 301 ha (3.01 km^2)
    annual soil loss = 301 x 1.56 x 1100 x 0.125 x 0.1
    = 5,870 ton/year or
    = 3,335 m^3/year, with a soil bulk density of 1.76 m^3/ton.
- dugout/valley tank: for an overall mean catchment area of 165 ha (1.65 km^2),
    annual soil loss = 165 x 1.56 x 1,100 x 0.125 x 0.1/1.76
    = 2,011 m^3/year.

These soil loss/sedimentation figures are compatible with the above estimated sedimentation rates.

Indicative figures for sediment loads related to erosion hazards in catchment areas and the size of catchment areas confirm the above estimates. (Erosion Hazard Map of Ghana, SRI, 1971 correlated with field observations by UNDP/ILO, 1986).

With an assumed overall moderate erosion hazard in the catchment, annual sedimentation loads according to these indicative figures would be:

- for dams with irrigation facilities: 1505 – 3010 m^3/year (average = 2260) and
- for dugouts and valley tanks: 825 – 1650 m^3/year (average = 1240).

For the project, the following figures were used to estimate the sediment deposit and the useful life of the project:

- dams with irrigation facilities: 3,300 m^3/year and 83,000 m^3 accumulated over a 25-year period
- dugouts/valley tanks: 2,000 m^3/year and 50,000 m^3 accumulated over a 25-year period

Most semi-arid areas have similar soil and climatic circumstances; it follows that siltation will occur naturally in all similar areas, sooner or later, depending on the level of exposure of the catchment area soil.

9.11.2 RECENT RESEARCH FINDINGS IN SEDIMENTATION

Sedimentation engineering is a very wide field in which much research has been done, but only the practical aspects are relevant here.

Erosion process takes this trend: weathering prepares the parent rock for erosion, raindrops dislodge the weathered materials, while run-off transports and eventually deposits them under different slope and channel configurations.

Contrary to the generally known scenario that soil conservation measures taken in a watershed generally reduce the rate of reservoir sedimentation, K. Mahmood (1987) argues that this is not so since only 10% of the materials eroded from a basin appears at the channel outlet, the rest is stored on hill slopes, in the valleys and within stream channels. He agrees however that the sediment load carried by a river is deposited in the reservoir because the transport capacity of flow diminishes with decreasing velocity.

The mitigating measures recommended, based on research findings, include:

a. control of sediment inflow into the reservoir
b. hydraulically remove the sediment load already deposited in the reservoir
c. dredging/desilting existing deposits and
d. construction of debris dam, i.e. first trap the sediment in an intermediate dam (referred to as debris dam) before discharging sediment-free flow into the reservoir! This will be an expensive remedy.

However, one needs not get involved in this academic exercise. In practice, the use of grass strips of about 30 m immediately upstream of the reservoir serves as an effective filter, allowing only relatively sediment-free flow into the reservoir (figures 6.1B and 6.2). And whatever percentage of sediment gets near the reservoir can be so trapped and so prolong the useful life of the reservoir.

9.11.3 MITIGATING MEASURES

The subject of watershed management (or land conservation or environmental protection as it is variously referred to), is essentially the adoption of soil conservation measures in catchment areas of dams and valley tanks to minimise soil loss due to erosion etc. Among the measures to build into the project design are:

1. reduction to the barest minimum of human and livestock activities within the catchment or watershed.
2. where the cultivation of the watershed cannot be avoided, proper

cultivation programmes should be worked out, which should include:
i. detailed Topographical Survey at scale 1:1,000,
ii. design of watershed farming conservation measures which should include contour farming, use of terraces, bunds, gully stabilisation and grassing depending on the situation
iii. ploughing to a maximum depth of 0.75 m to minimise soil loosening and
3. Agro-forestry. Contour farming reduces the amount of run-off by reducing the length of strips over which run-off has to flow, thus encourages infiltration and minimising soil movement at the same time.

Grassing, especially drought-resistant species like vetiver, bahama and paspallum, binds the soil together and when planted along the contour, also encourages water harvesting, which further increases infiltration, leads to higher crop yield and holds sediment back; this way, it is only sediment-free water that is allowed into the reservoir, which ultimately reduces the rate of reservoir siltation.

9.12 Involuntary Resettlement in Water Resources Development

9.12.1 INTRODUCTION

Displacement/dislocation of people and communities living on and/or with their means of livelihood situated on dam/reservoir sites/zones, as well as having rights of way for supply canal/pipelines, is one of the potential negative impacts of most water resources development projects – dams and reservoirs, irrigation and drainage schemes and hydropower generation. Such projects have the potential for and/or give rise to severe social, economic and environmental problems

Social problems:
- people are relocated to unfamiliar environments
- community/family linkages get weakened and dispersed
- cultural identity and traditional values/monuments are lost or diminished.

Economic problems:
- production systems get dismantled
- assets and sources of income are lost
- productive capacities often become less in new locations and competition for resources/clientele sometimes becomes greater and therefore less favourable.

To ensure minimum adverse impact of such involuntary resettlement on the resettlers and host communities as well as the environment, adequate conscious planning and resources must be built into and committed to the project from inception. The following should be considered and taken into account in water resources development that has potential for displacing people and communities:

9.12.2 THE OBJECTIVE OF RESETTLEMENT PROGRAMMES

The objective of involuntary resettlement programmes should be:

a. to ensure that the displaced population receives commensurate benefits (compensation) for its displacement
b. to ensure at least an equal standard of living with the original settlement
c. to form an integral part of the project and be dealt with appropriately from inception.

9.12.3 ISSUES TO CONSIDER IN RESETTLEMENT PLANNING AND IMPLEMENTATION

i. As much as possible, involuntary resettlement should be avoided or minimised, i.e. all viable alternatives should first be explored e.g. reducing dam height, realigning canal/pipeline may significantly reduce resettlement needs. Change of site can also be considered.

ii. Where dislocation/resettlement cannot be avoided, appropriate resettlement plans should be developed; resettlers should be provided with sufficient investment resources and opportunities to share the project benefits. They should also be:

   a. compensated for their losses at full replacement costs (cash/kind), prior to actual movement.
   b. assisted in moving and supported during the transition period on the new site.
   c. assisted in efforts to improve on their previous living standards, income-earning capacities etc.
   d. the poorest groups should receive the greatest attention.

iii. Beneficiary participation should be ensured in the planning and implementation of the resettlement scheme. Appropriate social patterns should be established; existing sociocultural institutions of resettlers and those of receiving communities (hosts) should be supported.

iv. Resettlers should be fully integrated socially and economically into host communities to minimise the adverse impacts on host communities. This is best done by locating resettlement sites within beneficiary areas and through consultation with future hosts.

v. Land, housing, roads, water supply, schools, health services and other compensations should be provided to the adversely affected population, indigenous people/groups, including nomads who have usufructuary and/or customary rights to the land and other resources taken up by the project.

The absence of legal land title by such groups should not bar them from adequate compensation.

9.12.4 RESETTLEMENT PLANNING

A detailed resettlement plan, timetable and budget are a sine qua non for a successful resettlement programme. The plan should be built around a strategy and package, aimed at improving or at least restoring the economic base of the people to be relocated.

Experience has shown that cash compensation alone is inadequate – very soon, it leaves the people more impoverished as the cash is usually never well utilised.

Preference should be given to land-based strategies for people dislocated from agricultural settings. But where suitable agricultural land is not available in areas/locations acceptable to the resettlers, non-land-based strategies, built around local or nearby opportunities for employment or self-employment could be used.

A suitable resettlement plan should provide for the following, among other things:

1. Identifiable stakeholders and their specific roles.
2. Beneficiary participation – resettlers and host communities.
3. Provision of shelter, roads, water supply, schools, health facilities etc, commensurate with the needs of the population being resettled with reasonable projection for population growth.
4. Provision of economic activities – agriculture, marketing, social services.
5. Environmental protection and management.
6. Access to training, employment or self-employment.
7. An implementation schedule synchronised with the project implementation plan e.g. resettlers must move and be reasonably settled before reservoir impounding commences.
8. Provision of adequate funds to ensure successful planning and implementation as well as monitoring and evaluation.

Chapter X
ECONOMICS OF WATER RESOURCES DEVELOPMENT

This subject will be treated in three parts for convenience and ease of comprehension:

a) defining the problem
b) planning for water resources development and
c) engineering economy in water resources planning.

10.1 The Engineer's Problem and the Need for Management

The world is expanding in population and urban-industrial development. This naturally puts an ever-increasing pressure on the available water resources potential. If the society is to survive, the need for a better-planned water resources management programme becomes imperative, especially in semi-arid areas of the world.

The engineer's role in water resources planning is to maintain a proper planning methodology, capable of producing a viable resource management programme.

He needs to make his designs flexible to accommodate future needs, the nature of which he can hardly predict or anticipate, especially in this rapidly changing world. But how can he devise a management system for adequately sensing changes in human needs (e.g. urban water supply) as they occur and quickly adjust this management policy and even design as necessary? The answer to this question will be found in the second part – Planning for Water Resources Development.

10.2 The Engineer's Attempt

10.2.1 THE RESOURCES TO BE MANAGED

The resource to be managed is essentially the water retained in each phase of the hydrological cycle (chapter I) as it continues its unending journey. Thus the hydrological cycle is the vast natural resource system, which provides the basic management input in water resources planning.

10.2.2 MODIFICATIONS

The water resources planner (engineer) seeks to modify this natural system by introducing structural measures that force the movement of water to places and times better suited to meeting known human needs, e.g.

a. He puts a barrier (dam) across a river, thus preventing the water from flowing away in the rainy season and making same available for use much later in the dry season.
b. Flood-control efforts are directed towards preventing flow into certain areas by means of hydraulic structures.
c. Irrigation and land drainage efforts are positive attempts at making water available as and when needed for crop production.
d. Water supply programmes seek to make water available for municipal, industrial and livestock uses by means of diversion and storage structures.

He also seeks to modify by non-structural measures, human activities so as to better conform to known movement patterns e.g.

    a.   Flood Control and Flood Plain Management.
    b.   Water supply systems to existing cities and development of new cities closer to available systems.

The best management, therefore, is one which adopts the best possible combination of measures.

10.3 The Approach to the Engineering Planning

Engineering planning attempts to assess the available water resource potential with a view to evolving the best development strategy that will optimise the use of the resource at minimum cost. The attempt, therefore, is to examine how best to apply the concepts of social welfare to specific engineering design choices.

The concept of optimum programme alone is used less in present-day planning because of the recognition of the need to consider additional factors, e.g. political wishes/programme of the ruling party.

While microeconomics examines the benefits and costs, and provides rules for maximising profits minus costs as a step to enhance the welfare of the general public, engineering economy provides the procedures for the cost analysis of alternatives to find the least-cost approach irrespective of viewpoint.

10.4 Planning for Water Resources Development

10.4.1 PLANNING CONCEPT, INVOLVEMENT AND METHODOLOGY

Planning is the orderly consideration of a project from the initial statement of purpose through the evaluation of alternatives to the final decision on a course of action. The purpose of the planning in the context of water resources development is to determine the development programme and management that will contribute most to the welfare of all the people – the would-be beneficiaries.

Planning includes design and all associated works, except detailed engineering design of the structures. It also includes the evaluation of alternatives by the principles of engineering economy discussed later in this chapter. It is the basis of the decision to proceed with or abandon a proposed project. Planning is therefore the most important aspect of the engineering of any project. However, because each project is unique in its physical and economic setting, there is no known single simple process of general applicability that will inevitably lead to the best decision i.e. there is no substitute for 'engineering judgement' in the selection of the best method of approach to project planning; but each individual step towards the final decision must be subjected to and supported by a quantitative analysis rather than estimates or judgements.

One final characteristic of a good planning methodology is that it should be flexible in response to changing social objectives, better and faster techniques of analysis and advancing technology for plan implementation.

10.4.2 LEVELS OF PLANNING

Planning can be at the national, regional and project levels, each of which has specific objectives. At the national level in most cases, the minister of water resources, assisted by the national council of water resources, is responsible for setting and specifying targets for water resources management. But for specific purposes, there are water resources development agencies, e.g. river basin development authorities, water boards and corporations, and the marines services etc, all responsible for different aspects of the national water resources development programme.

Although regional planning is generally acceptable for other things, a natural 'region' for water resources planning is the river basin (or watershed); see chapter IX.

Planning at the lowest level is planning for specific actions, it is at this level that important decisions are taken, regarding the effectiveness of a water management project. This is called the project planning stage, though a physical project may not necessarily emerge.

10.4.3 PHASES OF PLANNING

Generally, planning is done in phases for ease of analysis and to avoid unnecessary expenditure likely to result from non-phasing.

10.4.3.1 *Reconnaissance Study*
This can be described as a 'coarse screen' intended to eliminate projects or actions clearly infeasible without extensive study, i.e. to identify those activities which deserve further study.

10.4.3.2 *Feasibility Study*
This is usually a thorough evaluation of the feasibility of the proposed activity

in the process of which one formulates a description of the most desirable alternative (action), called the plan. In some cases, a single study is sufficient because the nature of the problem may be easily evaluated. In other cases however, one or more pre-feasibility studies are undertaken to examine various aspects of the proposals, the idea of these several sequential studies being to reduce the planning costs by testing the weakest aspects of the project first.

Feasibility study usually requires the specifications of structural elements in sufficient detail to permit an accurate cost estimate, and in the final stage of this phase, the details of design must be carefully examined and construction drawings and specifications produced.

10.5   Preparation for Planning

For the different phases and stages of planning for water resources development, a great deal of data is required. Prominent among the data required are climatological, stream, flood and water quality records, topographical and geological, population and economic data, soil maps, land capacity and land use data, regional economy, unemployment records, wage rates etc. All assumptions must be clearly and comprehensively stated. The data are then analysed and eventually used in project formulation which is the ultimate objective of planning. The planner should isolate the factors which are most important and concentrate his planning efforts on them first, before going on to others, e.g. irrigation in an arid region relies heavily on water, hence he should consider sources of water and other relevant issues first. His findings will determine whether to proceed with planning. It is important to note that data for planning must be current and of sufficient length of time, especially hydrological data. Otherwise, new data collection stations may have to be set up and further studies suspended in the meantime.

10.6   Planning Objectives

To say that a water project is feasible implies that it will effectively fulfil the objectives for which it is intended. To measure feasibility therefore, the project purposes and objectives must be specified before planning begins. Objectives differ with different levels of planning, for instance:

a. At the national level, the objectives of water resources planning are:
   i. formulate and implement a water resources development master plan
   ii. coordinate the development and utilisation of water resources for irrigation, water supply and hydro-power generation
   iii. promote/support and coordinate the study, development and improvement of the nations groundwater and surface water resources potentials etc., whereas

b. At the river basin level (regional), the overall objective is: to develop the water resources potentials of the catchment area for multi-purpose use, i.e. agriculture, municipal and industrial needs and for hydropower generation.

The objectives of a particular project will differ, even within a river basin objective. At every stage of planning, the planner should subject the objectives to all possible alternative measures or combinations of such measures. The objectives should remain broad until the tests eliminate some of the measures.

10.7 Projections for Planning

Projections are often based on assumptions of present-day demands and population growth rates. Possible changes in technology should be considered in projections of the future. The principles of demand and elasticity are applicable to water needs.

Water requirements are often estimated on the basis of assumed ultimate requirements when all land in the service area of, say, an irrigation project, is fully developed and requires water in maximum quantities, based on projections of population, industrial and agricultural needs, using water rates per person or per hectare as the case may be. Projections for Water Resources projects are generally for the next 50 to 100 years.

As an example of the projection strategy, let us consider the requirement for the Oshun River Basin in Nigeria for a target year, say, 50 years ahead. Based on an annual population growth rate of 2.5% and a present per capita consumption of 50 litres per day (lcd), projections of 80 lcd and 180 lcd for the next 30 and 20 successive years i.e. years 2033 and 2053 respectively, reckoning from year 2003. These projections would result in a Municipal and Industrial water demand of 1761 m cm for a population of 21.77 million. In a similar manner, using the appropriate evapotranspiration rates, crop, soil and other crop water requirement factors, an irrigation requirement of 273.7 m cm for a 35,000 ha irrigation project would be required for the year 2033; thus a total water requirement projection of 2034.7 m cm would be made for the entire basin for that year for these two consumptive water users. Livestock is not considered because the project area is not suitable for large-scale livestock farming. This projection seems feasible because there is an estimated mean annual run-off of 3,300 m cm in the basin, i.e. a surplus of at least 1,200 m cm (Tahal Nigeria Ltd., 1980). The surplus allows for poultry, rabbitery, piggery, which are becoming popular in the area.

Unrealistic values of water needs (requirements) should be avoided because they may lead to over-design and hence excessive costs and if supply is limited, it may lead to the diversion of water from other uses (or sources) which are also economic but may be of lower priority. A good example is the proposal to divert water from the Ogun Basin where there is a reliable surplus all through the year to the Oshun Basin where there is the possibility of unreliable dry season flows though there is a reasonable overall surplus as discussed above.

In the case of the Ogun Basin mentioned above, the original proposal for a dam in the basin, Ikere Gorge Dam, based on an over-estimation of the water requirements was to build up a reservoir capacity of 1,300 m cm at an estimated construction cost of over Naira 80 m in 1979. On closer study, it was discovered that a reservoir capacity of only 565 m cm was not only sufficient but in fact the only reasonable capacity to build up in view of the lower annual catchment run-off of only about 700 m cm. Thus the estimated project cost was reduced by this single exercise by about 50%.

10.8 Project Formulation

Once the basic data and future projections are assembled, actual formulation of the project can commence. It must be emphasised that this phase of project planning requires skill and imagination.

Compilation of a list of alternatives is the first important consideration. The planning process should then be an evaluation of all possible alternatives and competing uses with respect to project features and water use.

As project formulation proceeds, it may become evident that new data or projections are required or that some revision of the available background data is required.

Another important step in project formulation is to define boundary conditions which will restrict the extent of the project e.g.

a. A policy decision may reserve certain lands for specific purposes like parks and recreation
b. Available water may be limited or subject only to minor changes.

10.9 Project Evaluation

Once alternatives have been defined, the planner's task is to provide data which help in the choice among alternatives. For an economic objective, data on benefits and costs are required in addition to the technical ones required for technical feasibility. At this stage, all costs must be specified in sufficient details so that costs can be reliably estimated. All costs, including those induced by the project should be included. The methods used for the estimation of both costs and benefits are under Engineering Economy (10.13).

It should be noted that not all benefits and costs can be measured in terms of money; for instance, the social cost of requiring people to move away from a reservoir site (involuntary resettlement) or the peace of mind gained by the reduction of flood hazards must be dealt with in descriptive rather than monetary terms.

10.10 Environmental Considerations in Planning: Environmental Impact Study (E.I.S.)

Where projects seem essential and feasible, the planner will find it necessary to consider very carefully the ecological impact of the proposed project on the adjacent stream and areas. He then proceeds to attempt to develop a plan which will have minimal detrimental effects. A partial list of environmental consequences of water resource projects worth considering at this stage will include:

i. Degradation of downstream channel or coastal beaches by loss of sediment trapped in reservoir.
ii. Creation of barrier to normal migration routes of land animals (e.g. cattle) by a reservoir.
iii. Damage to fish by passage through pumps or turbines or over the spillways of high dams, thus reducing the yield expectations of fishermen downstream.
iv. Damage to stream bank vegetation by alteration of flow patterns in a stream.
v. Loss of unique geological, historical, archaeological or scenic sites flooded by a reservoir, (e.g. in the resettlement programme of the Oyan River Dam near Abeokuta in south-west Nigeria, the shrine of a traditional deity first had to be re-established in a colourful ceremony at the new village (resettlement site) before contractors were allowed to clear the old shrine, which fell within the reservoir of the dam.

A clear distinction should be made between temporary (due to construction operations, tree clearing etc.) and long-term effects due to the very existence of that project in the area in question. Water resources development and the environment is discussed in greater detail in chapter IX.

10.11 Some Common Pitfalls in Project Planning

Pitfalls may negate an otherwise excellent effort to make a sound project out of a planning exercise. In general, the only protection against these difficulties is to be constantly alert against incorporating them in the analysis. A few of these pitfalls are:

a. Failure to consider all alternatives.
This is the commonest pitfall, especially with respect to non-engineering alternatives – flood plain management may be overlooked by an engineering designer. Similarly, simple engineering alternatives are generally overlooked especially when they depart from the traditional solutions.

Example:

Problem: To reduce seepage through a small earth dam.

Alternative Solutions: Consulting soils engineers will recommend the construction of a clay cut-off wall and blanket on the upstream face instead of an environmentalist's or water engineer's preference for relief wells (through filters) on the downstream side which could provide a good source of filtered water as well as a seepage control measure.

b. A priori decisions.

These are decisions taken on certain project features before economic analysis and are never checked e.g. a management decision to use sprinkler irrigation systems on all fields because of the fear of high transmission and field application water losses.

c. Use of next-best alternative.

In the absence of any other estimate of benefits of a project it is not uncommon to use the cost of the next-best alternative as a measure of such benefits – this is based on the premise that the proposed project would have as much benefit as the cost of the alternative project if the original project were dropped.

The planner should check for these or other pitfalls and attempt to eliminate them from his planning strategy.

10.12 Multi-purpose Projects

Multi-purpose projects are those planned for multi-purpose benefits, e.g. irrigation, municipal and industrial needs and hydro-power generation.

10.12.1 PROBLEMS OF MULTI-PURPOSE PLANNING

The basic factor in design is to find a compromise among the competing objectives. A working plan must be devised which permits a reasonably efficient operation for each purpose although maximum efficiency is not necessarily attained for any single purpose. The unique factor of design is to select physical works and an operation plan which is an effective compromise among the various uses.

There are two possible extremes in the allocation of reservoir storage.

i. Assume that no storage is jointly used. In this case, the storage requirements of all uses are pyramided to create a large storage requirement for all functions, which can be economically attained when unit cost of storage is constant or decreases as total storage increases.

ii. Assume that all storage is jointly used. This assumption results in maximum economy since the required storage is no greater than that required for any one of the several purposes.

Situations described above are rare and the usual multi-purpose project is designed to fall somewhere between these two extremes.

10.12.2 FUNCTIONAL REQUIREMENTS OF MULTI-PURPOSE PROJECTS

The success which can be obtained in achieving joint use of storage space in a multi-purpose project depends on the extent to which the various purposes are compatible. We shall use a multi-purpose project for irrigation, water supply and hydro-electric power as an example.

i. Irrigation
Water requirements for irrigation are generally seasonal, having a maximum value at the middle of the dry season (January or February in Nigeria, July in Uganda etc.) and hardly any demand in the peak of the rains (July in Nigeria, May in Uganda etc.)

Since water requirements do not vary from year to year (except in extremely low rainfall years), and unless the project area is increased, the average annual demand will remain constant. And because irrigation is an insurance against crop failure resulting from drought, it is only reasonable to maintain as much reservoir storage as possible, consistent with known current demands.

ii. Water supply
Water requirement for municipal, industrial and livestock uses are more nearly constant throughout the year than irrigation requirements. However, a seasonal maximum is usually experienced in the dry season. Because of population increase, the demand normally increases gradually from year to year, hence, provision must be made for this increase. Adequate reservoir space must therefore be made to avoid water shortage during droughts.

iii. Hydro-electric power
Power demand generally has a marked seasonal variation, depending on the area served. Also, most power plants are parts of an interconnected system and considerable flexibility is therefore possible in coordinating power needs with other water uses. Similarly, because power production is not a consumptive use of water, it is more compatible with other uses. For instance, water released for irrigation use downstream may first be passed through a turbine to produce power as well, whereas in extreme situations, power production may be limited to times when releases are necessary for other uses or to discharge excess water. This combination of water demand for several uses is schematised in Figure 10.1. Note that although navigation is included in figure 10.1, releases for it have no noticeable effect on the annual demand curve. The consumptive uses which most of the dams discussed here will cover are water supply to humans, livestock and irrigation.

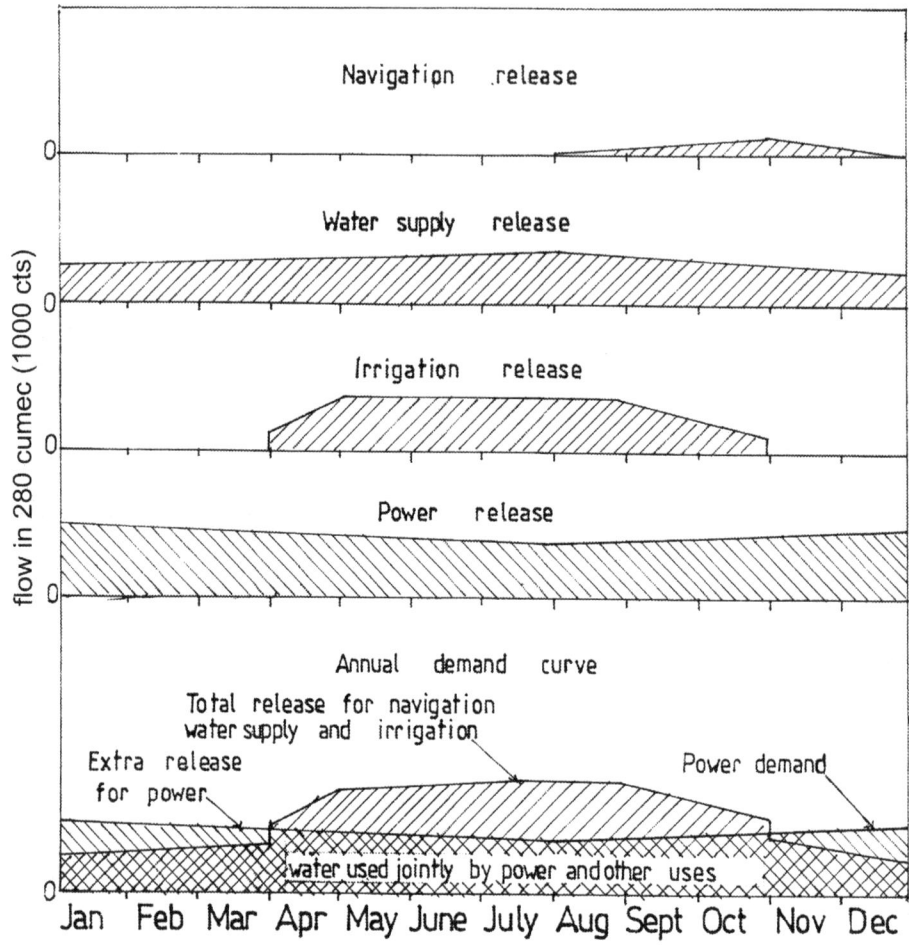

Figure 10.1: Water Demand for Several Uses in a Multipurpose Project

10.13 Engineering Economy in Water Resources Planning

10.13.1 INTRODUCTION

Engineering Economy is the Science of applying economic criteria to select the best of a group of alternative engineering designs. The relevant questions, which Engineering Economy sets to solve, are:

a. Why embark on this project?
b. Why do it now?
c. Why do it this way?

Most engineering problems have a set of major alternatives, each with its own set of sub-alternatives. However, at every stage of the analysis, the relative economy of these alternatives and sub-alternatives should be considered in making a rational choice.

Examples:
i. In the design of the Ikere Gorge Dam of the Ogun-Oshun River Basin Development Authority, Nigeria; there were two possible alternatives of diverting the river flow – either by means of a tunnel through the hillside or a conduit. Both alternatives were technically feasible but the adoption of the tunnel was technically more difficult, requiring the use of foreign expertise and special equipment. It therefore involved a rather high foreign currency component. In the final analysis, the conduit alternative was chosen because of ease of construction, the use of very little foreign expertise and equipment all resulting in a lower overall cost and lower foreign currency component from 12 to 7½% of the total project cost. (author's personal experience).
ii. There are two alternative ways of rehabilitating a silted up dam/reservoir – by desilting the reservoir or by raising the embankment. This is discussed conclusively in chapter VI.

10.13.2 SOCIAL IMPORTANCE OF ECONOMY IN WATER RESOURCES ENGINEERING DESIGN

Once completed, major water control structures can be altered only with difficulty or not at all. There are only relatively few suitable dam sites, most are marginal [like the Kyera Dam Site discussed in Section 6.3.4 and appendix 3] and once appropriated, the possibilities for economic multi-purpose development are very limited [if not altogether lost]. Once an irrigation project is developed, it cannot be moved, if unfavourable soil and/or climatic conditions are discovered later. There is a sobering finality in construction of a river basin development, and it behoves us to be sure we are right before we go ahead.

This excerpt from *A Water Policy for the American People*, The Report of the President's Water Resources Policy Commission, 1950, Vol. 1 p.18, summarises the social importance of sound economic decisions on major water resources projects.

10.13.3 STEPS IN ENGINEERING ECONOMY STUDY

The following is a list of steps usually taken in engineering economy studies:

a. Define in physical terms, each seemingly promising alternative.
b. Translate into money estimates as far as possible the physical estimates of each alternative especially for the items that may affect the choice between alternatives. These estimates should include dates and magnitudes of disbursement, lives and salvage values of structures and other assets.
c. Money estimates should be put in comparable units, using the appropriate (current) conversion factors e.g. US$ or Euros.
d. A definite choice (or recommendation) should be made from among the possible alternatives. This final choice is influenced by comparison in terms of money units and by other factors like the intangible (or irreducible) political consideration or even rural development policy pursuit.

Example of the application of the above procedure:
Two alternative plans are considered for a section of an aqueduct. Plan A uses a tunnel; plan B uses a section of lined canal and a section of steel flume. In plan A, the estimated first cost of the tunnel is US $450,000, its estimated annual maintenance cost is $4,000, and its estimated life is 100 years. Estimated first costs and lives for the elements of plan B are canal (excluding lining), $120,000, 100 years; canal lining, $50,000; flume, $90,000, 50 years. The annual maintenance cost is $10,000. The interest rate to be used in the economy study is 6% per annum. The study period is 100 years. All salvage values are assumed to be negligible. There are no estimated revenue differences between the two plans and no other cost differences are anticipated. (For instance there is no expected difference in water loss between the two alternatives.)

The comparison of equivalent annual cost for the two plans is as follows:

Plan A

Capital recovery cost for tunnel	= $450,000 x 0.06018	= $27,081
Annual maintenance cost		= 4,000
Total annual cost		= 31,081

Plan B $

Capital recovery cost for canal	= $ 120,000 x 0.06018	= 7,222
Capital recovery cost for canal lining	= $ 50,000 x 0.08718	= 4,359
Capital recovery cost for flume	= $ 90,000 x 0.06344	= 5,710
Annual maintenance cost		= 10,500
Total annual cost		= 27,791

In the comparison of Plans A & B, the first-cost figures for tunnel, canal lining and flume are multiplied by compound-interest factors that depend on their respective estimated lives and the 6% interest rate. The appropriate factor to convert an investment into an equivalent annual cost is designated as the capital recovery factor 1 and may be computed from the expression:

$[i(1+i)^N]/\{(1+i)^N - 1\}$
Where i = interest rate per annum (expressed as a decimal fraction) and
N = years of estimated life.

TABLE 10.1   CAPITAL RECOVERY FACTORS (f)

YEARS (N)	INTEREST RATE (i)						
	5	6	8	10	12	15	20
5	0.23097	0.23740	0.25046	0.26380	0.27741	0.29832	0.33438
10	0.12950	0.13587	0.14903	0.16275	0.17698	0.19925	0.23852
15	0.09634	0.10296	0.11683	0.13147	0.14682	0.17102	0.21388
20	0.08024	0.08718	0.10185	0.11746	0.13388	0.15976	0.20536
25	0.07095	0.07823	0.09368	0.11017	0.12750	0.15470	0.20212
30	0.06505	0.07265	0.08883	0.10608	0.12414	0.15230	0.20085
35	0.06107	0.06897	0.08580	0.10369	0.12232	0.15113	0.20034
40	0.05828	0.06646	0.08386	0.10226	0.12130	0.15056	0.20014
45	0.05626	0.06470	0.08259	0.10139	0.12074	0.15028	0.20005
50	0.05478	0.06344	0.08174	0.10086	0.12042	0.15014	0.20002
60	0.05283	0.06188	0.08080	0.10033	0.012013	0.15003	
70	0.05170	0.06103	0.08037	0.10013	0.12004	0.15000	
80	0.05103	0.06057	0.08017	0.10005			
90	0.05063	06032	0.08008	0.10002			
100	0.05038	0.06018	0.08004	0.10001			

f = A/P, where f = capital recovery factor
A = Annual payment
P = Present worth

Table 10.1 contains capital recovery factors for a number of values of i and N. When any present sum of money is multiplied by the capital recovery for N year and interest rate i, the product is an annual figure sufficient to pay exactly the present sum in N years with interest rate i. For example, the tabulated cost in plan B shows $4,359 as the capital recovery cost of $50,000, including 6% interest on each year's unpaid balance.

The comparison of the $31,081 annual cost of plan A with the $27,791 annual cost of plan B is representative of the innumerable annual cost comparisons that may be made in engineering economy studies. It is common for alternative designs to require different investments in fixed assets. In this example, the total investment figures are $450,000 and $260,000 (120,000 + 50,000 + 90,000) respectively. The extra investment of $190,000 (450,000 − 260,000) produces advantages that have a money value, such as lower maintenance costs and longer lives for the assets. In this instance, the annual cost comparison tells us that these advantages are insufficient to justify the extra investment. Plan B, therefore, should be selected unless plan A has some other advantages, so-called irreducibles or intangibles, that were not reflected in the estimates of costs and lives and that are believed to be sufficient to justify extra costs of $3,290 (31,081 − 27,791) a year.

Note that this comparison is valid only for the first 50 years of the structure, i.e. the estimated life of the flume. Thereafter, the flume may have to be replaced at an additional cost.

It should be emphasised that annual cost calculations of the type here illustrated are valid regardless of the scheme of financing to be employed. Even though it is convenient to explain the capital recovery factor in terms of a loan transaction, the computed totals for Plans A and B are the equivalent annual costs whether the entire first cost of the proposed assets is to be borrowed, whether it is to be entirely equity capital or whether some combination of borrowing and equity capital is contemplated. The preceding statement is subject to the qualification that the interest rate used in the economy study needs to be the rate that is most appropriate for the particular circumstances.

10.13.4 INTEREST RATES AND THE CHOICE OF PROJECT ALTERNATIVES

In the example discussed above, using an interest rate of 2% rather than 6% puts plan A at an advantage by a margin of about $4,760 per annum. In the same vein, an 8% interest rate puts plan B at an advantage by a margin of about $7,500 per annum. At an interest rate of about 4%+, the two plans have equal annual costs. It is thus clear that the choice of interest rate greatly influences the engineering design selected. At low interest rates (2–4%) many proposed investments, as in plan A, will appear economical whereas the same investments appear unduly costly at interest rates of 6–8%. The rate to use, therefore, should be the minimum attractive rate of return, all things considered. Rates commonly used in Nigeria are: a) commercial bank rates, and b) agricultural credit bank rates. These rates generally exceed 10%.

Where there is very high competition for capital as in competitive private enterprise, the use of relatively high interest rates (e.g. 10%) becomes necessary in economy studies – this has the effect of eliminating the least productive of the proposed investment opportunities and of conserving the limited funds for use in the most productive places.

In contrast to the foregoing, the practice in the United States of America is to use an investment rate equivalent to the base cost of borrowed money for public works, whereas other lending houses use a certain percentage above the minimum lending rate.

10.14 Estimated Lives of Hydraulic Structures

Although as mentioned under 10.7 above, the life expectancy of most hydraulic structures is generally put at between 50 and 100 years, the costs of public river basin development projects are usually computed on the basis of a 100-year life. For economy studies however, shorter periods than life may combine to reduce the project life expectancy considerably. A competitive private industry may use a pay-off period of 10 years for assets having expected life of 20 years or more. The economic factor of importance in this case is the capital recovery cost (or annual investment), which is the interest on investment plus depreciation (or amortisation). It is always proportional to investment, regardless of the way it is described.

10.15 Taxes and Other Investment Charges (Capital Recovery Costs)

These are charges which should be included in the overall project cost; for example, a dam of estimated 50-year life, using an interest rate of 6%, will be subject to investment charges of 6.34% of first cost. (Note that there is no tax on public works.) If the same dam were built on a privately owned facility, having an 8% cost of capital, 1.5% general property tax and income tax of 2.54% of first cost, investment charges would be 12.1%. This example shows that there is a difference in overall project cost, where taxes and other investment charges are involved.

10.16 Probability Concept in Planning

All projects are planned for the future whereas the planner is uncertain of the precise conditions to which the projects will be subjected. The structural designer knows the intended loads for the structure but he has no assurance that the loads will not be exceeded. This uncertainty is countered by making reasonable assumptions and allowing a generous factor of safety. The water resources engineer is less certain of the flows which will affect the project – he is also not certain of other events affecting design e.g. future water requirements, benefits and costs. However, a serious error in the estimate of expected hydrology can have very serious negative effects on the economy of the entire project. This is why economists use the term, 'ceteris paribus'.

Therefore, since the exact sequence of future stream flow cannot be predicted, the probability of such hydrological events has to be estimated. A detailed discussion of the statistical methods of estimating probability cannot be covered in this handbook, but the concept is applied in the succeeding portions of this chapter.

10.17 How Hydraulic Structures Should Be Designed

Many water resources design problems are related to infrequent extreme events e.g. extreme values of streamflow. At the time of design, the dates of such extreme events are unpredictable. The extreme values of such events can be predicted only by using the principle of relative frequency in the long run i.e. probability.

Case study

REQUIRED:

To determine whether to increase the most economic capacity of a spillway. Note that the greater the capacity of the spillway, the larger its first cost and hence the larger its annual investment charges. On the other hand, the greater the capacity, the less frequently will the spillway overflow (i.e. cost: benefit comparison). It is presumed that each overflow will cause some damage to property with resulting cost to the property owners.

OBJECTIVE:

To design a system that will minimise the sum of the annual costs of the increased capacity (investment charges, maintenance costs etc.) and the average annual cost of damage by an excessive overflow.

EXPERIENCE:

Experience shows that for many types of hydraulic structures, it turns out to be economical to design against extreme events so infrequent that they may not occur at all during the life of the structure e.g. a structure with an estimated life of 50 years might be designed to withstand a flood that is equalled or exceeded only once in 50 years. In this case it is not the average annual damages that matters but the annual cost of the risk of damage which is the product of the estimated probability that the event will be equalled or exceeded in any one year and the estimated cost of the adverse consequences if the event occurs. Table 10.2 illustrates an economic study based on this premise.

The table records an economy study for a large electric utility that acquired a small hydroelectric plant in connection with the purchase of a small utility property. It seemed to the engineers of the large company that spillway capacity of 48 cusec provided by the dam at the plant was inadequate; it was believed that a flood exceeding this capacity would overtop and wash out a section of the earth dam whose replacement cost was $400,000.

In order to determine what, if any, increase in spillway capacity was justified, the engineers estimated the cost of increasing the cost of the spillway capacity by various amounts. They also estimated the long-run frequencies of floods of various magnitudes, expressed as probabilities of occurrence in any year. Annual investment charges on the spillway enlargement was assumed as 10.38% (capital recovery based on a 50-year life with interest at 6%, property taxes at 1.5%, income taxes at 2.54%. (The annual 'cost of the risk' was calculated as the product of the expected damage of $400,000 if spillway capacity were exceeded and the estimated probability that the capacity would be exceeded in one year. The total annual costs are lowest for an investment of $62,000. This protects against a flood of 76 cusec, expected, on the average, once in 200 years.

In the majority of flood problems, damage is not constant but is an increasing function of flood magnitude. With a 48cusec spillway capacity, a 2000 cusec flow might cause $300,000 of damage while an 85 cusec flow might do $460,000 damage. In this case, the area under the damage versus probability curve must be determined.

Example:
Given the flow-probability and flow-damage data for the project (Table 10.3), find the average annual flood damage.

TABLE 10.2 ANNUAL COSTS FOR DIFFERENT PROPOSED SPILLWAY CAPACITIES (AFTER HAZEN)

(Figures in US $)

SPILLWAY CAPACITY (CUSEC)	PROBABILITY OF GREATER FLOOD IN ANY YEAR	COST OF ENLARGING SPILLWAY CAPACITY TO PROVIDE FOR THIS FLOOD	ANNUAL INVESTMENT CHARGES AT 10.3%	ANNUAL COST OF FLOOD RISK	SUM OF ANNUAL COSTS
1700	0.05	No cost	0	20,000	20,000
2000	0.02	30,000	3,114	8,000	11,000
2300	0.01	46,000	4,775	4,000	8,775
2700	0.005	62,000	6,436	2,000	8,436
3000	0.002	81,000	8,408	800	9,208
3300	0.001	104,000	10,795	400	11,195
3600	0.0005	103,000	13,494	200	13,694

TABLE 10.3 FLOW-PROBABILITY AND FLOW-DAMAGE DATA

(Figures in $)

Peak flow (cusec)	Probability of flow being equalled or exceeded in any year	Expected damage
1700	0.05	0
2000	0.02	200,000
2300	0.01	320,000
2700	0.005	400,000
3000	0.002	460,000
3300	0.001	500,000
3600	0.0005	540,000

Compute the cost of flood risk as follows:

Range of peak (cusec)	Average damage ($)	Probability of flow in interval	Annual damage ($)
1700–2000	100,000	0.05–0.02	3000
2001–2300	260,000	0.02–0.01	2600
2301–2700	360,000	0.01–0.005	1800
2701–3000	430,000	0.005–0.002	1290
3001–3300	480,000	0.002–0.001	480
3301–3600	520,000	0.001–0.0005	260
		Average annual damage	9430

The probability of occurrence of a flood peak between 76 and 85 cusec is 0.003. Since such a flood would cause about $430,000 damage, the average risk in any year is 0.003(430,000) = $1290. The total for all intervals, $9430, is the annual average flood damage. If a larger spillway were assumed, new estimates of damage would be required.

10.8 Economy Study for Private Versus Public Works

The monetary value assigned to an enterprise depends on whose viewpoint is primary, i.e. individual investor (or group), community in a specific area or the entire nation.

Thus for private companies and individuals, the effect of the expected project on the receipts and disbursements by the company is of paramount consideration – this view reflects the owner's interest and it is essential for the company or individual to survive in competing with others.

For government on the other hand, it is not sufficient to analyse the economy of the project only from the point of view of costs and revenues of the water corporation or board (for water supply) or river basin authority (for overall water resources development), but in terms of the overall (total) benefit to the public e.g. the expansion of a water works by government to effectively utilise the raw water potentials of a newly expanded dam should be seen in its effect on improved water supply for municipal and industrial purposes in terms of direct revenue to the water board. In both cases, the emphasis is on benefits to whom so ever they may accrue.

10.19 Benefit:Cost Analysis of Alternative Public Works Projects
(A CASE STUDY FROM NIGERIA)

The average annual damage from floods in a river basin is estimated to be Naira 400,000. Estimates are made for several alternative proposals for flood-mitigation works. Channel improvements would increase the capacity of the stream to carry stream discharge. There are two possible sites, A and B, for a dam and storage reservoir. Because dam site A is located in the reservoir area for B, only one or the other of these sites may be used, but not both. Either site may be used alone or may be combined with channel improvement. It is also possible to use channel improvement alone. Table 10.4 shows the estimated first cost of each project, the estimated annual damages due to floods with each project, the annual investment charges, using an interest rate of 3%, and the estimated annual disbursement for operation and maintenance. The life of the channel improvements is estimated as 100 years. The final column of the table gives the sum of annual damages and annual project costs. This sum is a minimum for project (III) in Table 10.4, the development at site B alone.

TABLE 10.4 ECONOMIC ANALYSIS OF PROPOSALS FOR FLOOD MITIGATION IN IBADAN, NIGERIA

PROJECT	INVESTMENT (NAIRA)	AVERAGE ANNUAL FLOOD DAMAGES (NAIRA)	ANNUAL INVESTMENT CHARGES (NAIRA)	ANNUAL OPERATION AND MAINTENANCE	SUM OF ANNUAL DAMAGE AND PROJECT COSTS, (NAIRA)
No flood mitigation at all	0	400,000	0	0	400,000
I. Channel improvement alone	500,000	250,000	28,720	100,000	378,720
II. Development at site A alone	3,000,000	190,000	94,950	60,000	344,950
III. Development at site B alone	4,000,000	125,000	126,600	80,000	331,600
IV. Site A with channel improvement	3,500,000	100,000	123,670	160,000	383,670
V. Site B with Channel improvement.	4,500,000	60,000	155,320	180,000	395,320

The more conventional way to analyse such public works proposals is by means of benefit:cost ratio. Table 10.5 illustrates such an analysis. The benefits (to the flood victims) are the estimated annual reductions in flood damages. The costs (to the government and therefore to the taxpayers) are the annual investment charges plus annual disbursements for operation and maintenance.

The five benefits:cost ratios of Table 10.5 do not in themselves provide enough information to make an economic choice among the five projects. To use the benefit:cost ratio as a sound basis for project formulation, additional calculations are necessary. The additional profits added by each separable increment of costs should be computed, and the ratio of the increments of benefits to the corresponding increments of cost should be determined. Such an analysis will lead to the selection of project III, just as in the analysis shown in Table 10.4. The preceding statement assumes that extra costs are justifiable whenever the resulting benefits exceed the extra costs, but are not justified if the resulting benefits are less than the extra costs. In other words, the most economical design is the one that gives the greatest excess of benefits over costs.

Thus project II adds benefits of Naira 60,000 over project I, whereas costs are increased by only Naira 26,230; the ratio of extra benefits to extra costs is 2:29. Similarly, the extra Naira 51,650 of costs of project III over project II are justified by increased benefits of Naira 65,000; the incremental benefits:cost ratio is 1.26. But projects IV and V are clearly uneconomical as compared with project III because the added benefits are considerably less than the extra costs required to produce the benefits.

Figure 10.2 shows the benefits and costs of a hypothetical project graphically. From point A to B, benefits exceed costs, i.e., the benefit:cost ratio exceeds one. The curve of benefits minus costs (i.e. net benefits) shows a maximum at C. Change in benefits/Change in cost <1. The maximum ratio of benefits to costs occurs at D, and this should be the limit of project size if it is desired to obtain a maximum rate of return on the investment. The increment from D to C is however, economic since the rate of return on the increment exceeds the minimum attractive rate of return.

TABLE 10.5 BENEFIT: COST ANALYSIS OF FLOOD MITIGATION PROPOSALS, IBADAN, NIGERIA

Project	Annual Benefit (Naira)	Annual Costs (Naira)	Benefit : Cost Ratio	Benefits Minus Cost (Naira)
I	150,000	128,720	1.17	21,280
II	210,000	154,950	1.36	55,050
III	275,000	206,600	1.33	68,400
IV	300,000	283,670	1.06	16,330
V	340,000	335,320	1.01	4,680.

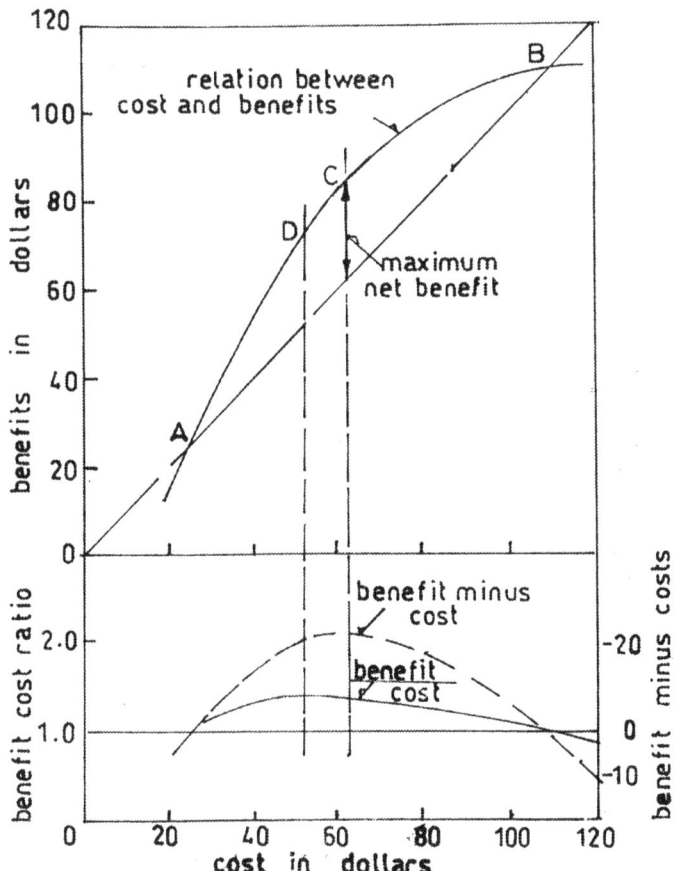

Figure 10.2: Benefit and Cost of a Hypothetical Project

As suggested in the previous paragraph, a project may be evaluated in terms of rate of return. Using the estimated cost and profit stream over the period of the project life, one may determine the rate of return on investment represented by the excess of benefits over costs. Projects may be ranked in merit in rate of return, and the decision to proceed can be based on a minimum acceptable rate of return as well as a benefit:cost ratio.

10.20 Capital Budgeting

Water resources project development is always allocated specific sums of money since it has to compete with other government projects like road development, education, defence etc. It is therefore obvious that no government agency, river basin authority or water board can ever execute all the water projects found feasible within its area of jurisdiction in any single year or even plan period.

Budgetary constraints may be in terms of limitation on the capital investment in one year or the net expenditure (i.e. investment + operation and maintenance – income to the funding agency). We may leave the details of why funds are short to the relevant ministry of finance but in general, the constraint will be applied as a means of rationing present funds and the limitation on construction costs only is most appropriate.

Because of the budgetary constraints, project analysis should serve to rank all possible projects in order of priority as those offering the highest return (or other government interests) can be built first. This principle of ranking enables an agency to obtain maximum returns on its investment. Note that governments find it difficult to reconcile such ranking with political pressures, which encourage uniform area distribution of projects rather than maximisation of returns. Intangible factors such as the need to stimulate the economy within a region also indicate a departure from the strict capital-budgeting rules.

10.21 Cost Allocation for Multi-purpose Projects

A multi-purpose project is one which is designed and operated to serve two or more purposes as discussed earlier. It is necessary to allocate the total cost among the several beneficiaries e.g. water supply, irrigation, power supply etc., to enable one to determine the contribution to cost-recovery by the different users.

There is no single method of general applicability that is foolproof and will yield unquestionably accurate allocations. Any method used must first set aside the separate costs that are specifically chargeable to a single function project, e.g. the Oyan River Dam Power Plant in South-west Nigeria alone cost about Naira 5.8m out of the overall cost of about Naira 72m. Other separable costs may be fish monk and ladder, water treatment intake etc.

The separable cost for a single function project is usually estimated as the total project cost less the estimated cost with that function omitted. In the case of Lekan Are Dam, near Abeokuta, Nigeria, the cost of the Water Treatment Intake could have been estimated as the total project cost (Naira 430,000) less the estimated cost without the treatment intake, if unknown (in this case, Naira 430,000 – 410,000 = 20,000).

The real problem of cost allocation is that of dividing the joint costs (total costs minus the sum of separable costs) among the project functions. Generally, two methods are considered applicable:

a. the remaining benefits method.
b. The Alternative Justifiable Expenditure method.

The application of these methods is illustrated in Table 10.6. The first line of the table presents the separable costs, totalling Naira 1,180,000. Since the total cost to be allocated is Naira 1,765,000, the joint cost is Naira 85,000.

TABLE 10.6 COST ALLOCATION FOR A MULTI-PURPOSE PROJECT COSTING NAIRA 1,765,000.*

LINE	ITEM	FLOOD MITIGATION	POWER	IRRIGATION	NAVIGATION	TOTAL
1	Separable costs	380	600	150	50	1180
2	Estimated profits	500	1500	350	100	2450
3	Alternate single-purpose cost	400	1000	600	80	2080

Remaining–benefits method

LINE	ITEM	FLOOD MITIGATION	POWER	IRRIGATION	NAVIGATION	TOTAL
4	Benefits limited by alternate cost	400	1000	350	80	1830
5	Remaining benefits	20	400	200	30	650
6	Allocated joint costs	18	360	180	27	585
7	Total allocation:					
	Naira	398	960	330	77	1765
	Per cent	22.5	54.4	18.7	4.4	100

Alternative Justifiable expenditure method

LINE	ITEM	FLOOD MITIGATION	POWER	IRRIGATION	NAVIGATION	TOTAL
8	Alternate cost less separable cost	20	400	450	30	900
9	Allocated joint cost	13	260	292	20	585
10	Total allocation:					
	Naira (N)	393	860	442	70	1765
	Percent (%)	22.2	48.7	25.0	4.0	100

*All cost figures in the table are $\times 10^3$

In the remaining benefits method, the joint costs are assumed to be distributed in accordance with the differences between the separable costs (line 1) and the estimated benefits of each function (line 2). In no case, however, are the benefits assumed to be greater than the cost of an alternate single-purpose project, which would provide equivalent benefits. Thus the remaining benefits (line 5) are the differences between the lesser value of lines 2 or 3 and the separable costs in line 1. The joint costs are distributed in proportion to these remaining profits added to the corresponding separable costs to obtain the total allocation (line 7).

Under the alternative justifiable expenditure method, the joint costs are assumed to be distributed in accordance with the differences between the separable costs and the estimated cost of a single-purpose project which would provide equivalent services and would itself be economically justifiable. These costs (line 8) are the differences between lines 1 and 3. The distributed joint costs (line 9) are added to the separable cost to obtain the total allocated cost (line 10). It should be noted that a large percentage difference in the allocation of joint costs is only a moderately small percentage difference in the allocated total cost because the separable costs are normally a large part of the total.

The greatest difficulty with both methods is that of estimating the benefits or alternative costs. From a practical viewpoint both methods give results which are within the probable limits of accuracy obtainable in any allocation.

In order to establish a pricing policy, it is sometimes necessary to allocate costs between several classes of users. When a long pipeline is constructed, it may be proper to charge users near the head of the line less for water than is charged to users at the far end of the line. In this case the allocation is often based on the proportional use of facilities – but this is not possible in practice.

No single method of cost allocation can be properly described as the best method. For establishing prices or allocating charges to various beneficiaries, any method, which is agreeable to all concerned, can be considered acceptable. Because of the arbitrary nature of cost allocation, it must be viewed as an accounting device totally unrelated to the economic evaluation of a project. Costs allocated to a specific project purpose should not be compared with expected benefits to justify the inclusion of the particular purpose in the project plans. If the remaining benefits method is used, it will automatically make each purpose appear to have a benefit–cost ratio in excess of one (Table 10.6).

A more simplified approach for small schemes with cost recovery objectives in mind was adopted in the Upper East Region of Ghana (IFAD-funded Land Conservation and Smallholder Rehabilitation Project). The option adopted was: 'allocate costs on the basis of the amount of water that goes to each measurable consumptive benefit area and even out by approximations.'

On that project, an average rehabilitated dam has a reservoir storage (capacity) of 265,000 m^3. The allocation is as follows:

i. Evaporation, Dead storage and Seepage
    (i.e. losses)   – 40%
ii. Livestock }
iii. Domestics }
iv. Irrigation } shared what remained in
    ratio: 1.2:1:4.4   – 60%

Based on this sharing ratio, the analysis is:
- The losses which amount to 40% are shared among the three benefit areas in the ratio of their benefits, reduced to 15, 18 and 67% respectively.
- Evaporation losses were estimated, based on the available scanty data and experiences elsewhere. Similarly, Domestic requirements were estimated. It was understood also that irrigation might extract more or less than estimated at the completion of the project, more so when the data on reservoir volumes were rather suspect; fish farming and Social Benefits were also left out.

Based on these uncertainties, it was assumed that 67–75 % of the common costs could be and were allocated to Irrigation, while other benefit areas shared 25–33%.

Using the project cost of $4,468,300, Irrigation-related costs were:
Option 1: 70% = US $3,127,810
Option 2: 67% = US $2,993,761

Going further to cost per ha, on a total Project hectarage of 555, the cost per ha of the two options are:

Option 1 = $5,636 and
Option 2 = $5,394

This approach implies that the losses and costs of other benefits were taken as sunk costs by the Project Authorities, while the beneficiaries of irrigation facilities were expected to pay the irrigation-related costs in form of cost recovery.

The option chosen was left to the Policy Makers, based on the crops grown under Irrigation and Government's Cost Recovery Policy.

REFERENCES

Chapter I

Wilson, E M, *Engineering Hydrology*, 3rd ed., Macmillan, reprinted 1985
USDA, *The Year Book of Agriculture – Water*, 1955
USDI, Bureau of Reclamation, *Design of Small Dams* – A Water Resources Technical Publication, 3rd ed. 1987
Jackson, I J, *Climate, Water and Agriculture in The Tropics*, 2nd ed., Longman, 1989
Raudkiri, I J, *Hydrology, An Advanced Introduction to Hydrological Processes and Modelling*, Pergamon Press, 1979

Chapter II

Faniran, A, 'Which Way to Effective Water Supply in Developing Countries?' Paper presented at the National Workshop on the 5th National Development Plan, University of Ibadan, Nigeria, March 1985
Nissen-Petersen, Erik & Lee, Michael, *Harvesting Rainwater in Semi-Arid Africa, Manual No 1 Water Tanks with Guttering and Hand-Pump*, Nairobi, Kenya, 1990
Uganda National Study in Support of the Inter-governmental Negotiation Committee on Drought and Desertification, MAAIF, 1993

Chapter III

Linsley, R K, M A Kholer and J L H Paulhus, *Hydrology for Engineers*, S.I. Metric Edition, McGraw-Hill Book Coy, London, 1988
Nilsson, Ake, *Groundwater Dams for Small-scale Water Supply*, IT Publications, 1988
Wilson, E M, *Engineering Hydrology*, 3rd ed., Macmillan, reprinted 1985
Wagner, E G & Lanoix, *Water Supply for Rural Areas and Small Communities*, World Health Organisation (WHO) Nomograph Series 42, 1959

Chapter IV

Gichuki, F N, *Utilization and Conservation of Wetlands – An Agricultural Drainage Perspective*, Department of Agricultural Engineering, University of Nairobi, 1992
Fatokun, Jide, 'Reservoir Structures for Rural Water Supply', Paper Presented at The Nigerian National Water Resources Institute, Kaduna, April 1987

Linsley, R K, M A Kholer and J L H Paulhus, *Hydrology for Engineers*, S.I. Metric Edition, McGraw-Hill Book Coy, London, 1988

Wilson, E.M., *Engineering Hydrology*, 3rd ed., reprinted 1985, Macmillan

USDI, Bureau of Reclamation, *Design of Small Dams* – A Water Resources Technical Publication, 3rd ed., 1987

Wagner, E G & Lanoix, *Water Supply for Rural Areas and Small Communities*, World Health Organisation (WHO) Nomograph Series 42, 1959

Chapter V

Award Consultants, February 1999 – *Kanyaryeru Resettlement Scheme Water Supply Feasibility Study Report*, submitted to K-2 Consult, Kampala, Uganda

Fatokun, Jide, 'The Water Component, South-west Region Agricultural Rehabilitation Project (SWRARP)', Report Submitted to MAAIF, Uganda, July 1995

——, Jide, March 1996, 'The Water Component, Livestock Services Project (LSP)', main report submitted to MAAIF, Uganda

——, Jide, 'Design Report LSP'

——, Jide, *Personal Experience in Hydrological & Feasibility Studies and Design of Water Projects in Uganda, 1994–2003*

Gaf Consult/Norconsult AB, Feb. 1993, 'The Study of Water for Livestock and Domestic Use and the Related Socio-economic and Environmental Issues in Lake Mburo National Park and Environs', report submitted to the Water Department, Ministry of Energy, Minerals and Environment Protection, the Republic of Uganda

Linsley, R K, M A Kholer and J L H Paulhus, *Hydrology for Engineers*, S.I., Metric Edition, McGraw-Hill Book Coy, London, 1988

Linsley, R K, M A Kholer and J L H Paulhus, *Engineering Hydrology*

Scott Wilson, Kirkpatrick & Partners, 'Iseyin-Oke-Iho Water Supply Scheme Hydrological Studies', report submitted to the water corporation of Oyo State, Nigeria, July 1977

Tahal Consultants (Nig.) Ltd., 'Ilorin Water Supply Extension Planning Report' for the Kwara State (Nigeria) Water Corporation, 1973

Chapter VI

Fatokun, Jide, 'Revised Design Report, The Water Component, Livestock Services Project', MAAIF, Uganda, 1999

——, Jide, 'Irrigation Planning and Development in Nigeria', in *Irrigation Development Planning*, edited by J R Rydzewski, Southampton University, 1977

Award Consultants, 'Kanyaryeru Resettlement Scheme Water Supply Feasibility Study', report submitted to K-2 Consult, Kampala, Uganda, February 1999

USDI, Bureau of Reclamation, *Design of Small Dams* – A Water Resources Technical Publication, 3rd ed., 1987

Ogun-Oshun River Basin Dev. Auth, Eng. Dept, 'Design and Construction of Small Earth Dams', paper presented at the 8th meeting of General Managers of RBDA's, Abeokuta, Nigeria, 1983

FAO, 'Effective Rainfall, Irrigation and Drainage Paper 25', 1974

Chapter VII

Fatokun, Jide, 'Construction Report, LSP, the Water Component', report submitted to MAAIF, Uganda, February 1999

MAAIF, 'Livestock Services Project, Bid Analysis and Evaluation', report submitted to the World Bank and Government of Uganda, 1997

Federation Internationale de Ingenieurs Conseils (FIDIC) – (International Federation of Engineers), 'Conditions of Contract For Works of Civil Engineering Construction, Part I' – General Conditions with Forms of Tender and Agreement; 4th ed., reprinted 1988 with editorial amendments, 1987

The World Bank, 'Project Supervision, Operational Directive (OD) 13.05', 1989

Chapter VIII

Biswas, A K, *Methodology for M & E of Integrated Land and Water Development, in Strategies for River Basin Management – Environmental Integration of Land and Water in a River Basin*, D. Reidel Pub. Coy, Dordrecht, Holland, 1985

International Fund for Agricultural Development (IFAD), Monitoring and Evaluation Guidelines for Irrigation Projects; Rural Development M&E Series No 3, 1989

Lundqvist, Jan, Lohn Ulrik & Falkenmark, Malin Eds., *Strategies for River Basin Management*, D. Reidel Pub. Coy, 1985

USDI, Bureau of Reclamation, *Design of Small Dams* – A Water Resources Technical Publication, 3rd ed., 1987

Chapter IX

Faniran, A, 'Planning in Developing Countries – Reality, Harmony and Environment. Third World Planning Review', 1989, pp175–188

Fatokun, J O, 'The Effects of Angle of Slope and Vegetative Cover on Soil Erosion by Raindrop Impact', unpublished B.Sc. Thesis, University of Ife, Nigeria, 1971

Gichuki, F N, 'Utilization and Conservation of Wetlands – An Agricultural Drainage Perspective', Department of Agricultural Engineering, University of Nairobi, Kenya, 1992

Hamil, A T & Asiamah, R D, Soil Erosion in the Savannah Zones of Ghana, Kwadaso Soil Erosion Research at Bawku-Manga Research Station

Mahmood, K, 'Reservoir Sedimentation', World Bank Technical Paper Number 71, 1987

The World Bank, Involuntary Resettlement, OD. 4.30, June 1990

The World Bank, Environment Department, Environmental Assessment Source Book Volume II – Sectoral Guidelines, December 1991

UNDP/ILO, *Small Earth Dams*, Training Guide and Technical Guide for Special Public Works Programme Workers, 1986

Vohra, B B, *Problems Related to Coordinated Control and Management of Land and Water Resources*, some perceptions derived from the Indian Experience, in Strategies for River Basin Management, D.R. Pub. Coy., 1985

Chapter X

Are, Lekan & Fatokun, Jide, 'Rural Water Supply – the Ogun-Oshun River Basin Development Authority Experiences', a Second National Workshop on International Drinking Water Supply and Sanitation Decade Paper, Owerri, Nigeria, February 1983

Fatokun, Jide, 'Economics of Water Resources Planning', paper presented at the conference on Strategies for the fifth National Development Plan, 1986–1990, organised by the Nigerian Institute of Social and Economic Research (NISER), Ibadan, and the Federal Ministry of National Planning, Lagos, held at the University of Ibadan, November, 1984

James, L Douglas & Lee, Robert R, *Economics of Water Resources Planning*, McGraw-Hill, 1971

Linsley, Ray K & Franzini, Joseph B, *Water Resources Engineering*, 3rd ed, McGraw-Hill, 1979

Ogun-Oshun River Basin Dev. Auth, Eng Dept, 'Design and Construction of Small Earth Dams', paper presented at the 8th meeting of General Managers of RBDA's, Abeokuta, Nigeria, 1983

Rydzewski, J R, *Project Formulation, Appraisal and Monitoring in Irrigation Development Planning* (with special reference to conditions in Africa, South of the Sahara), ed. by J W Rydzewski, Southampton University, 1977

Tahal Consultants (Nig.) Ltd. & Associated Engineers and Consultants (Nig.) Ltd., 'Oshun River Basin – Master Plan for Development of Water Resources', final report, submitted to Ogun-Oshun RBDA, Abeokuta, Nigeria, 1980

Tahal Consultants Ltd., Oyan River Dam, 'Design Report', submitted to Ogun-Oshun RBDA, Nigeria, 1980

UK Ministry of Overseas Development, *A Guide to Project Appraisal in Developing Countries*, HMSO, London, 1972

Federal Ministry of Water Resources, Garki, Abuja, Nigeria, *The Wealth of Water – 1st Anniversary*

Appendix I
DEFINITIONS

agriculture: the science (or art) or practice of cultivating the soil and rearing animals with the ultimate objective of providing food, clothing and cash crops for the agriculturist.

agronomy: the science of soil management and crop production, i.e. the crop production aspect of agriculture.

animal husbandry: the science of breeding and caring for farm animals.

aquifer: water-bearing soil strata; it could be unconsolidated materials like sand, gravel and glacial drift or consolidated materials like sandstone and limestone in wide joints and solution passages.

artesian wells: wells drilled into confined aquifers.

as-built drawings: prior to the issuance of the completion certificate, the contractor is obliged to submit to the employer a set of drawings indicating the exact locations of all aspects of the works as well as their finished dimensions. Such drawings are called 'as-built drawings'. This is generally very useful in carrying out maintenance works in the future.

benching: a construction device on an existing embankment (or hill slope) by which short reaches of level ground are created to enable the addition and compaction of fresh materials to bond properly into the existing embankment. This eliminates interface problems which could lead to seepage (leakage).

catchment area: also called the watershed, it is the area of land which contributes to the run-off that becomes available at the dam site; it is not only the size of the area that matters, the totality of what are called the catchment characteristics is equally important. Some of these characteristics are length of the valley up to the dam site, slope, nature of soil and vegetative cover.

coffer dam: a temporary dam constructed to divert stream flow away from construction area to enable the substantive construction to go on; it is usually necessary when constructing a dam on a perennial river or during the rainy season on an annual stream.

contingency: the allowance usually made in pricing works to cover such items as changes in plans due to unforeseen site conditions and/or omissions; it varies from 10% to 30% for earthworks.

crop: the produce of cultivated plants, especially cereals.

cultivate: prepare and use soil etc. for crops or gardening (break up the ground with a cultivator to produce a suitable tilth for infiltration and crop growth; also raise or produce crops.

dam: a barrier or structure (embankment) built across a water course (valley) to store water rather than let it flow away.

effective rainfall (or net rainfall {precipitation}): the rain remaining as run-off (run-off analysis) after all losses by evaporation, interception and infiltration have been allowed for.

equivalence of kind To facilitate cost comparisons, it is desirable to relate the values of alternative choices to a common unit e.g. tonnes of steel pipes, so many kilowatt hours of electricity; reduction of the hazards of flood damages will be easy to compare when reduced to money terms, i.e. equivalence of kind.

equivalence of time Projects executed at different times cannot be compared unless reduced to the same time period – equivalence of time. An earlier realisation of investment returns is desirable to the investor because it gives him greater flexibility for future actions.

forestry: agriculture incorporating the cultivation and conservation of trees.

gravity delivery: the method of water delivery from its source to the point of use by the natural force created by differences in elevation; water flows under gravity from a higher to a lower level. This is possible from a dam reservoir if the system is designed to take advantage of level differences on the surface of the land along the valley being dammed.

groundwater table (phreatic surface): the surface of saturation of an aquifer. Its stability depends on supply from above – high in wet and low in dry season. Groundwater moves gradually towards rivers/lakes: it comes out to the surface where there is an impervious layer below the aquifer where an outcrop brings the impervious layer near the surface.

An impervious layer overlies confined aquifer and so the aquifer is under pressure. Piezometric Surface is the height to which the water in a confined aquifer would rise if it could.

impound: collect or store water, hence, a dam reservoir impounds water and is sometimes called an impounding reservoir.

infiltration: the movement of water (run-off) from the surface of the earth into the soil; its rate depends on the soil's physical properties and the amount of time available before overland flow takes the water beyond the point of infiltration.

intangible (or irreducible)factors:
These are the aspects of projects (cost or benefit) not easy to quantify in terms of money e.g. the need to stimulate the economy in a given region (or state). It is of interest to note that in some cases, intangible factors may control the choice between competing alternatives.

percolation: the movement of water within the soil immediately following infiltration. It is responsible for water movement from the upper layers of the soil to the saturated zone above the groundwater table.

permeability: the property of an aquifer to allow water to pass through it. Aquifers with large pores, e.g. gravel, have high permeability; those with small pores like clay have low permeability.

reservoir: in relation to a dam, a reservoir is the 'bowl' that holds the water which the dam prevents from flowing away.

run-off: surface run-off is the water flowing on the surface of the earth, arising from the combination of excess volume above what the soil can absorb (infiltration) and slope which brings the accumulated water under the force of gravity until it discharges into a larger body, i.e. stream, river, lake and ultimately the ocean.

seepage: slow percolation of water through a dam, usually occasioned by weaknesses within the embankment.

small dams: dams not higher than 15 m from the lowest level in the river bed or valley.

sunk costs (or past expenditure):
These are costs incurred on a project prior to the current analysis. Since an engineering economy study needs analyse only the differences (i.e. the differences likely to affect the overall costs), sunk costs should be disregarded. Sunk costs should not have any influence on deciding among alternatives except as they affect future cash flows.

valley tank: a reservoir created by excavating soil within a valley for the purpose of storing water for use after the flood level goes down. It holds water below the surrounding ground surface

	and, because of this fact, the water requires some means of bringing it to the surface, manually or mechanically (pumping).
variations:	departure from the planned work, usually as a result of information becoming available after design and cost estimates.
weir:	a low dam built across a river to raise the water level upstream for pumping or diversion purposes or to regulate its flow into downstream areas; the former purpose would be more applicable in this monograph.

Appendix II

DAM COSTS AS AFFECTED BY CHOICE OF EMBANKMENT SLOPES, USING KENWA DAM AS AN EXAMPLE

NOTES

1. Calculation of quantities for upstream slope 2:1 is also contained in Appendix 4 for completeness.
2. The contractor's rate used for computing costs is Sh. 6,000 per m^3.

KENWA DAM, MBARARA DISTRICT – CALCULATION OF EMBANKMENT QUANTITIES: DOWNSTREAM SLOPE (D/S) = 2:1; UPSTREAM SLOPE (U/S) = 2:1

1	2	3	4	5	6	7	8	9	10	11	12	13
CHAIN-AGE (M)	INTERVAL L_1-L_2 (M)	ELEVATION (M)	HEIGHT H= $EL_{MAX} - EL_0$ (M)	CORE AREA 4(H-1) (M^2)	EMBANK-MENT AREA H(4 + 2H) (M^2)	SHELL AREA $COL_6 - COL_5$ (M^2)	AVERAGE CORE AREA (M^2)	AVERAGE SHELL AREA (M^2)	CORE VOL $COL_8 \times COL_2$ (M^3)	CUMU-LATIVE CORE VOL (M^3)	SHELL VOL $COL_9 \times COL_2$ (M^3)	CUMU-LATIVE SHELL VOL (M^3)
34		50	0	0	0	–						
40	6	47	3	8.0	30	22	4.0	11	24	24	66	66
50	10	46.9	3.1	8.4	31.62	23.22	4.2	22.61	42	66	226.1	292.1
60	10	48.4	1.6	2.4	11.52	9.12	5.4	16.17	54	120	161.7	453.8
70	10	46.9	3.1	8.4	31.62	23.22	5.4	16.17	54	174	161.7	615.5
80	10	45.8	4.2	12.8	52.08	39.28	10.6	31.25	106	280	312.5	928
90	10	44.7	5.3	17.2	77.38	60.18	15	49.73	150	430	497.3	1425.3
100	10	44.1	5.9	19.2	93.22	73.62	18.4	66.9	184	614	669	2094.3
110	10	44.0	6.0	20.0	96	76	19.8	24.81	198	812	748.1	2842.4
120	10	44.0	6.0	20.0	96	76	20	76	200	1012	760	2842.4
130	10	43.8	6.2	20.8	101.68	80.88	20.4	78.48	204	1216	784.4	3602.4
	10						20.5	79.65	205	1528	796.5	4386.8

1 CHAIN-AGE (M)	2 INTERVAL L_1-L_2 (M)	3 ELEVATION (M)	4 HEIGHT H = EL_{MAX} − EL_0 (M)	5 CORE AREA 4(H-1) (M^2)	6 EMBANK-MENT AREA H(4 + 2H) (M^2)	7 SHELL AREA COL_6 − COL_5 (M^2)	8 AVERAGE CORE AREA (M^2)	9 AVERAGE SHELL AREA (M^2)	10 CORE VOL COL_8 X COL_2 (M^3)	11 CUMU-LATIVE CORE VOL (M^3)	12 SHELL VOL COL_9 X COL_2 (M^3)	13 CUMU-LATIVE SHELL VOL (M^3)
140		43.9	6.1	20.4	98.82	78.42						
	10						19.8	74.85	198	1719	748.5	5183.3
150		44.2	5.8	19.2	90.48	71.28						
	10						18.8	69	188	1907	690	5931.8
160		44.4	5.6	18.4	85.12	66.72						
	10						15.4	52.17	154	2161	521.7	7143.5
170		45.9	4.1	12.4	50.02	37.62						
	10						10.2	29.81	102	2263	298.1	7441.6
180		47.0	3.0	8.0	30	22						
	10						4.4	14.44	44	2307	144.4	7586
190		48.8	1.2	0.8	7.68	6.88						
	8						0.4	3.44	3.2	2310	27.52	7613.5
198		50	0	-								
	-								Add core foundation	443		
									Total	2753		7614

KENWA DAM – CALCULATION OF EMBANKMENT SLOPE AREA: DOWNSTREAM SLOPE (D/S) = 2:1;   UPSTREAM SLOPE (U/S) = 2:1

1	2	3	4	5	6	7	8
CHAINAGE (M)	INTERVAL L_1-L_0 (M)	ELEVATION (M)	HEIGHT $EL_{MAX} - EL_0$ (M)	SLOPE LENGTH $\sqrt{h^2 + (2h)^2}$ (M)	AVERAGE SLOPE LENGTH (M)	SLOPE AREA(M^2) $COL_6 \times COL_2$	CULULATIVE SLOPE AREA(M^2)
31.5		51	0	0			
40	8.5	47	4.0	7.96	3.98	33.83	33.83
50	10	46.9	4.1	8.16	8.06	80.6	114.43
60	10	48.4	2.6	5.17	6.67	66.7	181.13
70	10	46.9	4.1	8.16	6.67	66.7	247.83
80	10	45.8	5.2	10.35	9.26	92.6	340.43
90	10	44.7	6.3	12.53	11.44	114.4	454.83
100	10	44.1	6.9	13.73	13.13	131.3	586.13
110	10	44.0	7.0	13.93	13.83	138.3	724.43
120	10	44.0	7.0	13.93	13.93	139.3	863.73
130	10	43.8	7.2	14.33	14.13	141.3	1005.05
140	10	43.9	7.1	14.13	14.23	142.3	1147.33
	10				13.83	138.3	1285.63

1	2	3	4	5	6	7	8
CHAINAGE (M)	INTERVAL L_1-L_0 (M)	ELEVATION (M)	HEIGHT $EL_{MAX} - EL_0$ (M)	SLOPE LENGTH $\sqrt{h^2 + (2h)^2}$ (M)	AVERAGE SLOPE LENGTH (M)	SLOPE AREA(M^2) $COL_6 \times COL_2$	CULULATIVE SLOPE AREA(M^2)
150		44.2	6.8	13.53			
160	10	44.4	6.6	13.13	13.33	133.3	1418.93
170	10	45.9	5.1	10.15	11.64	116.4	1535.33
172	2	47.2	3.8	7.56	8.86	17.72	1553.05
180	8	47.0	4.0	7.96	7.76	60.08	1613.13
190	6.5	46.8	4.2	8.36	8.16	53.04	1666.17
200	3.5	48.8	2.2	4.38	6.37	22.295	1688.465
205	10	50.3	0.7	1.26	2.82	28.2	1716.665
210	10	51.0	0	0	0.63	6.3	1722.965
	5				-	-	1723
				U/S =	1723		
				D/S =	1723		
				Both	3446		
				Filter belt reduce to	10,000 13,646 – save 288		

KENWA DAM, MBARARA DISTRICT – EMBANKMENT QUANTITIES (SHELL ONLY) U/S SLOPE = 2.5:1, D/S = 2:1

1	2	3	4	5	6	7	8	9	10	11	12	13
CHAIN-AGE (M)	INTERVAL L_1-L_2 (M)	ELEVATION (M)	HEIGHT H = $EL_{MAX} - EL_0$ (M)	CORE AREA 4(H-1) (M^2)	EMBANK-MENT AREA H (4 + 2H) (M^2)	SHELL AREA $COL_6 - COL_5$ (M^2)	AVERAGE CORE AREA (M^2)	AVERAGE SHELL AREA (M^2)	CORE VOL $COL_8 \times COL_2$ (M^3)	CUMU-LATIVE CORE VOL (M^3)	SHELL VOL $COL_9 \times COL_2$ (M^3)	CUMU-LATIVE SHELL VOL (M^3)
34		50	0	0	0	-		-			-	-
	6							11.93			71.58	71.58
40		47	3	8.0	32.25	23.85						
	10							24.84			248.4	319.98
50		46.9	3.1	8.4	34.23	25.83						
	10							17.80			178.0	497.98
60		48.4	1.6	2.4	12.16	9.76						
	10							17.80			178.0	675.98
70		46.9	3.1	8.4	34.23	25.83						
	10							34.76			347.6	1023.58
80		45.8	4.2	12.8	56.49	43.69						
	10							55.45			554.5	1578.08
90		44.7	5.3	17.2	84.40	67.2						
	10							74.96			749.6	2327.68
100		44.1	5.9	19.2	101.92	82.72						
	10							83.86			838.6	3166.28
110		44.0	6.0	20.0	105	85						
	10							85			850	4016.28
120		44.0	6.0	200	105	85						
	10							87.75			877.5	4893.78
130		43.8	6.2	20.8	111.29	90.49						
	10							89.11			891.1	5784.88

1 CHAIN-AGE (M)	2 INTERVAL L_1-L_2 (M)	3 ELEVATION (M)	4 HEIGHT $H = EL_{MAX} - EL_0$ (M)	5 CORE AREA $4(H-1)$ (M^2)	6 EMBANK-MENT AREA H (4 + 2H) (M^2)	7 SHELL AREA $COL_6 - COL_5$ (M^2)	8 AVERAGE CORE AREA (M^2)	9 AVERAGE SHELL AREA (M^2)	10 CORE VOL $COL_8 \times COL_2$ (M^3)	11 CUMU-LATIVE CORE VOL (M^3)	12 SHELL VOL $COL_9 \times COL_2$ (M^3)	13 CUMU-LATIVE SHELL VOL (M^3)
140		43.9	6.1	20.4	108.12	87.72						
150	10	44.2	5.8	19.2	98.89	79.69		83.71			837.1	6621.98
160	10	44.4	5.6	18.4	92.96	74.56		77.13			771.3	7393.28
170	10	45.9	4.1	12.4	54.22	41.82		58.19			581.9	7975.18
180	10	47.0	3.0	8.0	34.23	26.23		34.03			340.3	8315.48
190	10	48.8	1.2	0.8	8.04	7.24		16.74			167.4	8482.88
198	8	50	0									
	–											

Shell material for slope 2.5:1   = 8,482.9 m³
Shell material for slope 2:1   = 7,614 m³ (see separate calculation for slope 2:1)
Therefore, additional material required for slope 2.5:1 in excess of slope 2:1   = 8,482.9 – 7,614
                                      = 868.9 m³ or an increase of 11.4% or additional sh. 5.214 million

KENWA DAM – CALCULATION OF EMBANKMENT SLOPE AREA FOR U/S = 2.5:1, D/S SLOPE = 2:1

1	2	3	4	5	6	7	8
CHAINAGE (M)	INTERVAL L_1-L_0 (M)	ELEVATION (M)	HEIGHT $EL_{MAX} - EL_0$	SLOPE LENGTH $\sqrt{h^2 + (2h)^2}$ (M)	AVERAGE SLOPE LENGTH (M)	SLOPE AREA (M^2) $COL_6 \times COL_2$	CULULATIVE SLOPE AREA (M^2)
31.5		51	0	10.77			
	8.5				10.91	92.74	92.74
40		47	4.0	11.04			
	10				8.77	87.7	180.44
50		46.9	4.1	6.5			
	10				8.77	87.7	268.14
60		48.4	2.6	11.04			
	10				12.02	120.2	388.34
70		46.9	4.1	13.0			
	10				14.38	143.8	532.14
80		45.8	5.2	15.75			
	10				17.17	171.7	703.84
90		44.7	6.3	18.58			
	10				18.72	182.7	891.04
100		44.1	6.9	18.85			
	10				18.85	188.5	1097.54
110		44.0	7.0	18.85			
	10				19.12	191.2	1270.74
120		44.0	7.0	19.39			
	10				19.26	192.6	1463.34
130		43.8	7.2	19.12			
	10				18.72	187.2	1650.54
140		43.9	7.1	18.31			
	10				17.05	170.5	1820.54

1	2	3	4	5	6	7	8
CHAINAGE (M)	INTERVAL L_1-L_0 (M)	ELEVATION (M)	HEIGHT $EL_{MAX} - EL_0$	SLOPE LENGTH $\sqrt{h^2 + (2h)^2}$ (M)	AVERAGE SLOPE LENGTH (M)	SLOPE AREA (M^2) $COL_6 \times COL_2$	CULULATIVE SLOPE AREA (M^2)
150		44.2	6.8	16.50			
	10				14.63	146.3	1966.84
160		44.4	6.6	12.75			
	10				11.13	111.3	2078.14
170		45.9	5.1	9.50			
	2				10.14	101.4	2179.54
172		47.2	3.8	10.77			
	8				11.04	110.4	2289.94
180		47.0	4.0	11.31			
	6.5				8.62	86.2	2376.14
186.5		46.8	4.2	5.92			
	3.5				3.9	39.0	2415.14
190		48.8	2.2	1.88			
	10				1.57		
200		50.3	0.7	1.26			
	10				0.63		
210		51.0	0	0			
	5						
215							

Total for slope 2.5:1 = 2,415.14 m²
Total for slope 2:1 = 1,723 m² (see separate calculation for slope 2:1)
Therefore, excess for 2.5: over 2:1 = 692 m² or 40% more than slope 2:1
This generates additional costs of:
Slope trimming = Sh. 3.1 m
Grassing = Sh. 0.69 m
TOTAL = Sh. 3.79 m

KENWA DAM, MBARARA DISTRICT – EMBANKMENT QUANTITIES (SHELL ONLY) U/S SLOPE = 3:1, D/S = 2:1

1	2	3	4	5	6	7	8	9	10	11	12	13
CHAIN-AGE (m)	INTERVAL L_1–Col_5 (m)	ELEVAT-ION (m)	HEIGHT $h = El_{max} - El_0$ (m)	CORE AREA $4(h-1)$ (m^2)	EMBANK-MENT AREA $h(4+2h)$ (m^2)	SHELL AREA $Col_6 - Col_5$ (m^2)	AVERAGE CORE AREA (m^2)	AVERAGE SHELL AREA (m^2)	CORE VOL Col_8 x Col_2 (m^3)	CUMU-LATIVE CORE VOL (m^3)	SHELL VOL Col_9 x Col_2 (m^3)	CUMU-LATIVE SHELL VOL (m^3)
34		50	0	0	0	-	-	-			-	-
	6											
40		47	3	8.0	34.5	26.5		13.25			79.50	79.50
	10											
50		46.9	3.1	8.4	36.43	28.03		27.27			272.7	352.2
	10											
60		48.4	1.6	2.4	12.8	10.4		19.22			192.2	544.4
	10											
70		46.9	3.1	8.4	36.43	28.03		19.22			192.2	736.6
	10											
80		45.8	4.2	12.8	60.9	48.10		38.07			380.7	1117.3
	10											
90		44.7	5.3	17.2	91.43	74.23		61.17			611.7	1729.0
	10											
100		44.1	5.9	19.2	110.63	91.43		82.83			828.3	2557.3
	10											
110		44.0	6.0	20.0	114.0	94.0		92.72			927.2	3484.5
	10											
120		44.0	6.0	20.0	114.0	94.0		94			940	4424.5
	10											
130		43.8	6.2	20.8	120.9	100.1		97.05			970.5	5395
	10							98.57			985.7	6380.7

1 CHAIN-AGE (m)	2 INTERVAL L_1-Col_5 (m)	3 ELEVAT-ION (m)	4 HEIGHT $h = El_{max} - El_0$ (m)	5 CORE AREA $4(h-1)$ (m^2)	6 EMBANK-MENT AREA $h(4 + 2h)$ (m^2)	7 SHELL AREA $Col_6 - Col_5$ (m^2)	8 AVERAGE CORE AREA (m^2)	9 AVERAGE SHELL AREA (m^2)	10 CORE VOL Col_8 x Col_2 (m^3)	11 CUMU-LATIVE CORE VOL (m^3)	12 SHELL VOL Col_9 x Col_2 (m^3)	13 CUMU-LATIVE SHELL VOL (m^3)
140		43.9	6.1	20.4	117.43	97.03						
	10							92.57			925.7	7306.4
150		44.2	5.8	19.2	107.3	88.1						
	10							85.25			852.5	8158.9
160		44.4	5.6	18.4	100.8	82.4						
	10							64.22			642.2	8801.1
170		45.9	4.1	12.4	58.43	46.03						
	10							36.27			362.7	9163.8
180		47.0	3.0	8.0	34.5	26.50						
	10							17.05			170.5	9334.3
190		48.8	1.2	0.8	8.4	7.6						
	8											
198		50	0									
	—											

Shell material for slope 3:1 = 9,334.3 m^3
Shell material for slope 2:1 = 7,614 m^3 (see separate calculation for slope 2:1)
Therefore, additional material required
for slope 3:1 in excess of slope 2:1 = 9,334.3 − 7,614
= 1,720.3 m^3 or an increase of 22.6% or additional sh. 10.322 million

KENWA DAM – CALCULATION OF EMBANKMENT SLOPE AREA FOR U/S = 3:1, D/S SLOPE = 2:1

1	2	3	4	5	6	7	8
CHAINAGE (M)	INTERVAL $L_1 - L_0$ (M)	ELEVATION (M)	HEIGHT $EL_{MAX} - EL_0$ (M)	SLOPE LENGTH $\sqrt{h^2 + (2h)^2}$ (M)	AVERAGE SLOPE LENGTH (M)	SLOPE AREA (M^2) $COL_6 \times COL_2$	CULULATIVE SLOPE AREA (M^2)
31.5		51	0				
40	8.5	47	4.0	12.65	12.81	108.89	108.89
				12.97			
50	10	46.9	4.1	7.8	10.39	103.9	212.79
60	10	48.4	2.6	12.97	10.39	103.9	316.69
70	10	46.9	4.1	16.44	12.02	147.1	463.79
80	10	45.8	5.2	19.92	14.71	181.8	645.59
90	10	44.7	6.3	21.82	18.18	208.7	854.29
100	10	44.1	6.9	22.14	20.87	219.8	1074.09
110	10	44.0	7.0	22.14	21.98	221.4	1295.49
120	10	44.0	7.0	22.77	22.14	224.6	1520.09
130	10	43.8	7.2	22.45	22.46	226.1	1746.19
140	10	43.9	7.1	21.50	22.61	219.8	1965.99
	10				21.98	211.9	2177.89

1	2	3	4	5	6	7	8
CHAINAGE (M)	INTERVAL $L_1 - L_0$ (M)	ELEVATION (M)	HEIGHT $EL_{MAX} - EL_0$ (M)	SLOPE LENGTH $\sqrt{h^2 + (2h)^2}$ (M)	AVERAGE SLOPE LENGTH (M)	SLOPE AREA (M²) $COL_6 \times COL_2$	CULULATIVE SLOPE AREA (M²)
150		44.2	6.8	20.87			
160	10	44.4	6.6	16.13	21.19	185.0	2362.89
170	10	45.9	5.1	12.02	10.5	140.8	2503.69
172	2	47.2	3.8	12.65	14.08	123.4	2627.09
180	8	47.0	4.0	13.28	12.34	129.7	2756.79
186.5	6.5	46.8	4.2	6.96	12.97	101.2	2858.99
190	3.5	48.8	2.2	2.10	10.12	45.3	2903.29
200	10	50.3	0.7	0	4.53		
210	10	51.0	0	0	0		
215	5						

Total for slope 3:1 = 2,903.29 m²
Total for slope 2:1 = 1,723 m² (see separate calculation for slope 2:1)
Therefore, excess for 3.1: over 2:1 = 1180.29 m² or 68.5% more than slope 2:1
This generates additional costs of:
Slope trimming = Sh. 5.3 m
Grassing = Sh. 1.2 m
TOTAL = Sh. 6.5 m

Appendix III
DESILTING VERSUS EMBANKMENT RAISING IN DAM REHABILITATION

This appendix contains the following:

1. KISHANGURA DAM
   i. Site Selection Report
   ii. Area-Capacity Curves (Figure: Appendix 3.1)
   iii. Spillway Longitudinal Profile (Figure: Appendix 3.2)
   iv. Calculation of Embankment and Spillway Quantities

2. DYANGOMA DAM
   i. Longitudinal Profile of Dam Axis (Figure: Appendix 3.3)
   ii. Area-Capacity Curves (Figure: Appendix 3.4)

3. RWAMAKARA DAM
   - Area-Capacity Curves (Figure: Appendix 3.5)

4. KYERA DAM
   - Longitudinal profile of Dam Axis (Figure: Appendix 3.6)

Kishangura Dam – Detailed Site Inspection/Selection Report

SITE:	KISHANGURA	STATUS:	DAM REHABILITATION
COUNTY:	NYABUSHOZI	S/COUNTY:	KINONI
DISTRICT:	MBARARA	DATE:	AUGUST 1995

Description of Physiography:

The existing embankment requires reshaping; the water in the reservoir has been reduced to within about 60 m of the embankment. The reservoir was said to have been virtually full on a continuous basis until about 1989 when the rains had been most unreliable. The interceptor drain excavated to intercept the run-off and its load of silt from the rather steep right bank needs reshaping and slope re-establishment; stone pitching will be required at its entrance into the reservoir.

The observed extensive silting is as a result of the free movement of human and livestock populations within the reservoir.

Water Resources Potential

The catchment seems to be capable of yielding reasonable run-off in a good rainfall year. The luxuriant vegetation in the upper reaches of the reservoir and the fact that there was still a good volume of water in the reservoir (though heavily infested by aquatic plants) indicate that the run-off is generally good. The water retention capacity of the reservoir is also adjudged to be good.

Design Water Requirements

The populations to be served were estimated as follows:
- human: 4,000
- cattle: 20,000
- sheep/goats: 1,000

Land Issues

It is an existing facility situated on public land; there is therefore no land ownership problem.

Disposition of Beneficiaries

Their full participation was said to be guaranteed, but should be taken up further before rehabilitation works begin.

Additional Design Information Required

Topographical survey should extend from 200 m downstream of embankment to 1 km upstream. It should conform with the terms of reference prepared for this purpose. Because extensive desilting by 0.5–1.5 m may be required, the present water level should be clearly indicated in relation to the downstream area for gravity delivery purposes.

Notes on Embankment Quantities

See the alternative calculations for desilting and embankment raising options in chapter VI of this book.

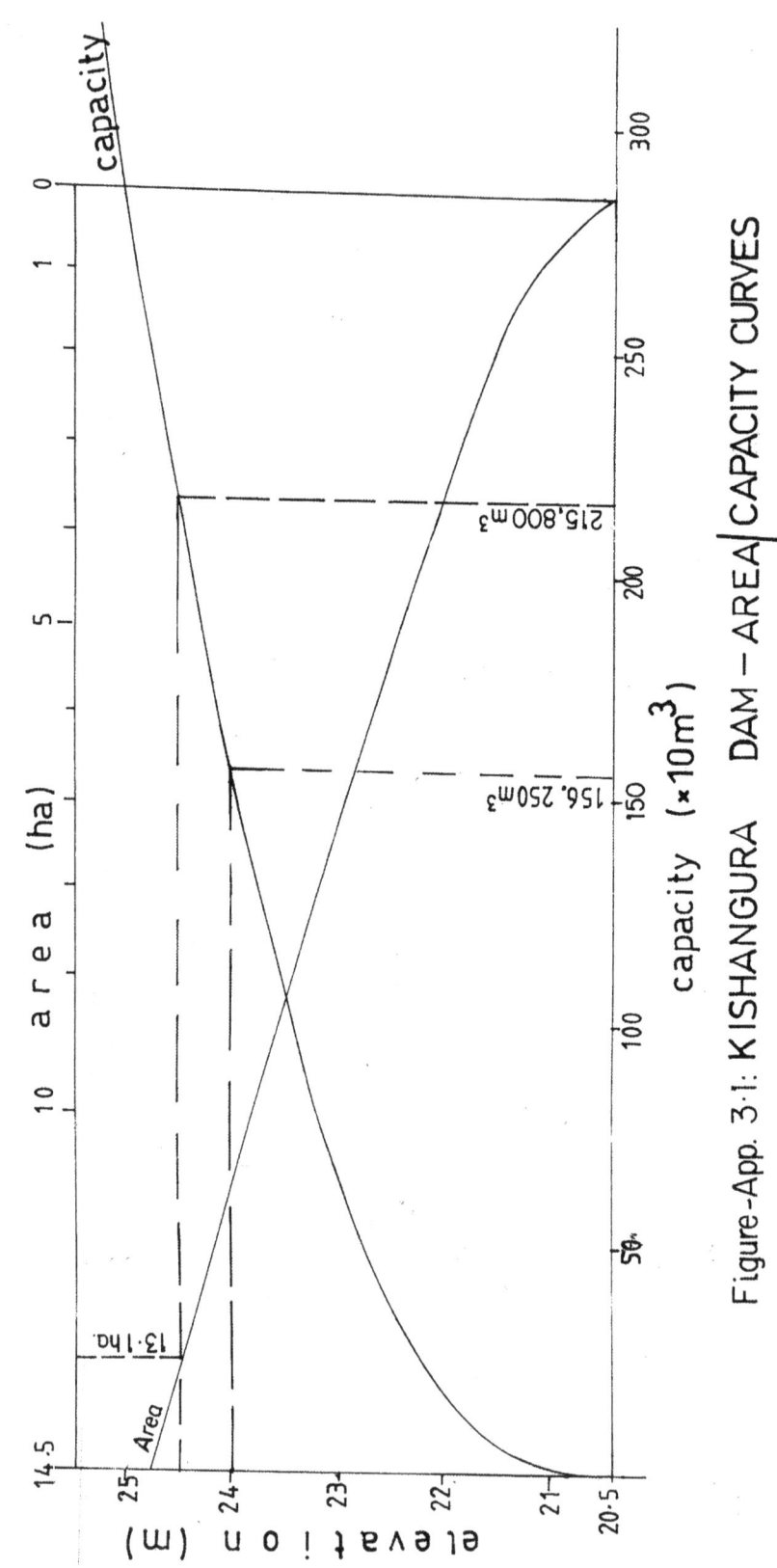

Figure-App. 3.1: KISHANGURA DAM – AREA/CAPACITY CURVES

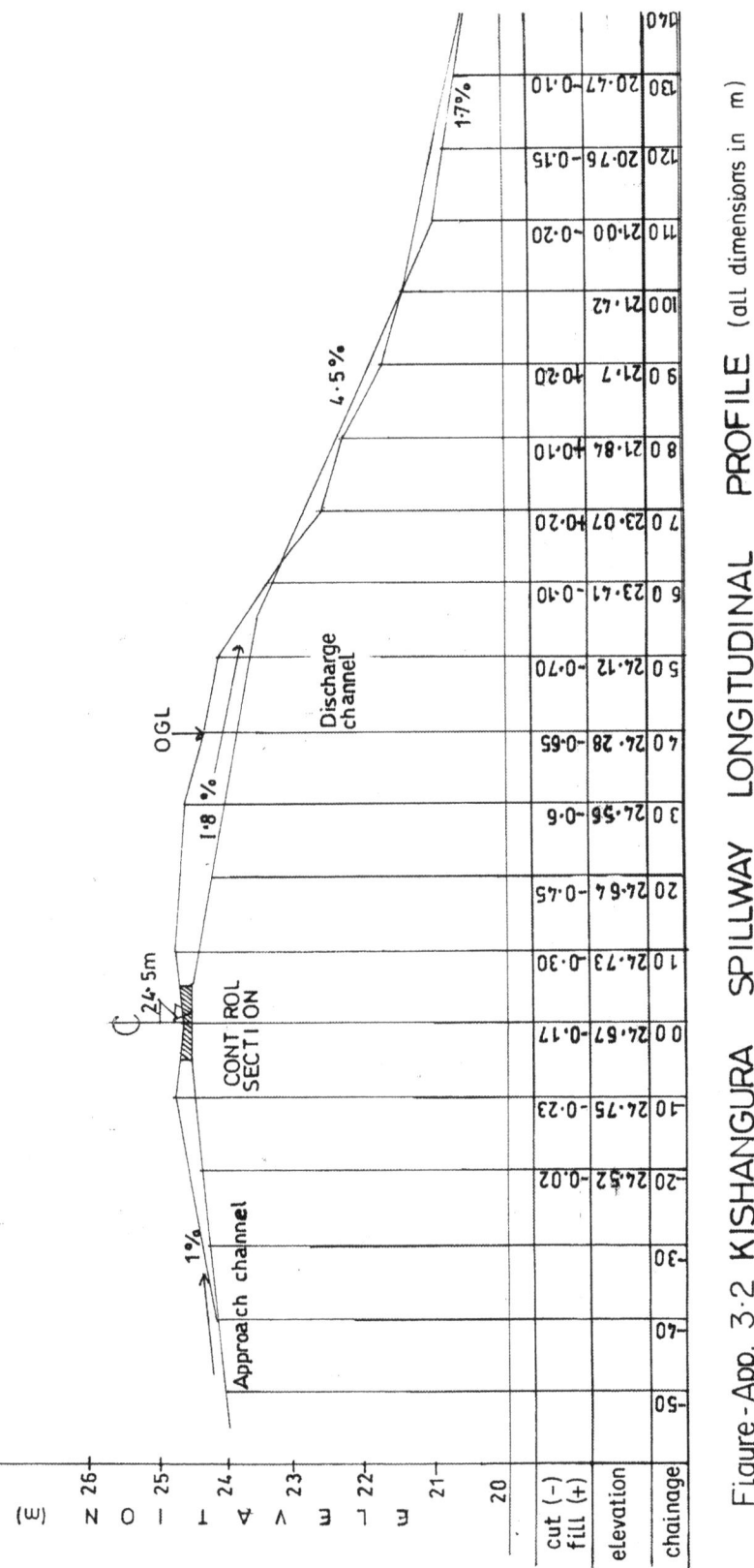

Figure-App. 3.2 KISHANGURA SPILLWAY LONGITUDINAL PROFILE (all dimensions in m)

Kishangura Dam: Calculation Of Spillway Excavation Quantities

Approach

1. $h = 1.2$, top width $= 23.6$
   $A = 354; B = 300; \sqrt{AB} = 325.88$
   $V = 391.95 \text{m}^3$

2. $h = 0.95$, top width $= 22.85$
   $A = 685.5; B = 600; \sqrt{AB} = 641.33$
   $V = 610.16 \text{m}^3$

3. $h = 0.48$, top width $= 21.45$
   $A = 1072.5; B = 1,000; \sqrt{AB} = 1,035.62$
   $V = 497.30 \text{ m}^3$

Approach Total $= 1,499.41 \text{m}^3$

Control

$h = 1.15$; top width $= 23.45$
   $A = 234.5; B = 200; \sqrt{AB} = 216.56$
   $V = 249.58 \text{ m}^3$

Discharge

1. $h = 0.9$, top width $= 22.7$
   $A = 1021.5; B = 900; \sqrt{AB} = 958.83$
   $V_1 = 846.10 \text{ m}^3$

2. $h = 0.6$, top width $= 21.8$
   $A = 1308; B = 1200; \sqrt{AB} = 1252.84$
   $V_2 = 752.17 \text{m}^3$

3. $h = 0.27$, top width $= 20.70$
   $A = 2288; B = 2200; \sqrt{AB} = 2243.57$
   $V_3 = 605.84$

Discharge Total $= 2204 \text{ m}^3$;
Spillway Total $= 3953 \text{ m}^3$

KISHANGURA DAM REHABILITATION: EXISTING EMBANKMENT QUANTITIES (1)

1	2	3	4	5	6	7
CHAINAGE	INTERVAL (M)	HEIGHT (M)	EMB. AREA H(4+2H) (M^2)	AVERAGE EMB AREA (M^2)	EMB. VOL COL5 X COL2 (M^3)	CUMULATIVE EMB VOL (M^3)
0		0.7	3.78			
10	10	2.12	17.47	10.63	106.3	106.3
20	10	2.62	24.21	20.84	208.4	314.7
30	10	2.91	28.58	26.40	264	578.7
40	10	2.75	26.13	27.36	273.6	852.3
50	10	2.83	27.34	26.74	267.4	1119.7
60	10	3.61	40.50	33.92	339.2	1458.9
70	10	3.7	42.18	41.34	413.4	1872.3
80	10	3.83	44.66	43.42	434.2	2306.3
90	10	3.56	39.59	42.13	421.3	2727.8
100	10	3.24	33.96	36.78	367.8	3095.6
110	10	2.59	23.78	28.87	288.7	3384.3
	10			20.13	201.3	3585.6

1	2	3	4	5	6	7
120	10	2.04	16.48	12.91	129.1	3714.7
130	10	1.38	9.33	10.17	101.7	3816.4
140	10	1.55	11.01	7.61	76.1	3892.5
150	10	0.76	4.20	2.4	24	3916.5
160	10	0.14	0.60	0.3	3	3319.5
170	10	2.0	s/w			
180						

KISHANGURA EMBANKMENT QUANTITIES – AS DESIGNED (2) EMB. CREST = ELE 25.5; NWL = ELE 4.5

1	2	3	4	5	6	7
CHAINAGE	INTERVAL	HEIGHT (M)	EMB. AREA H(4+2H) (M^2)	AVERAGE EMB AREA (M^2)	EMB. VOL COL5 X COL2 (M^3)	CUMULATIVE EMB VOL (M^3)
0		0	0			
	7			5.88	41.16	41.16
7		1.56	11.75			
	10			18.94	189.4	230.56
17		2.75	26.13			
	10			28.63	286.3	516.86
27		3.07	31.13			
	10			33.66	336.6	853.46
37		3.37	36.19			
	10			35.25	352.5	1205.96
47		3.26	34.3			
	10			36.05	360.05	1566.46
57		3.47	37.96			
	10			44.20	442.0	2008.46
67		4.12	5.43			
	10			49.52	495.2	2503.66
77		4.03	48.6			
	10			49.52	495.2	2998.86
87		4.12	50.43			
	10			49.52	495.2	3494.06
97		4.03	48.6			
	10			46.25	462.5	3956.56
107		3.79	43.89			
	10			38.42	384.2	4340.76

117		3.18	32.95			4634.46
	10			29.17	291.7	
127		2.7	25.38			4851.6
	10			21.87	218.7	
137		2.19	18.35			5056.86
	10			20.57	205.7	
147		2.52	22.78			5236.96
	10			18.01	180.1	
157		1.76	13.24			5320.76
	10			8.38	83.8	= 5321
167		0.66	3.51			

Additional shell material = 5,321 − 3,320 = 2,001
say 2,000 m^3

KISHANGURA SLOPES – AS DESIGNED (3)

1	2	3	4	5	6	7
CHAINAGE	INTERVAL (M)	HEIGHT (M)	SLOPE LENGTH $\sqrt{h^2 + (2h)^2}$ (M)	AVERAGE SLOPE LENGTH (M)	SLOPE AREA COL.5 X COL.2 (M^2)	CUMULATIVE SLOPE AREA (M^2)
0	0	0	0			
	7			1.75	12.25	12.25
7		1.56	3.49			
	10			4.82	48.2	60.45
17		2.75	6.15			
	10			6.51	65.1	125.55
27		3.07	6.87			
	10			7.21	72.1	197.65
37		3.37	7.54			
	10			7.42	74.2	271.85
47		3.26	7.29			
	10			7.53	75.3	347.15
57		3.47	7.76			
	10			8.49	84.9	432.05
67		4.12	9.21			
	10			9.11	91.1	523.15
77		4.03	9.01			
	10			9.11	91.1	614.25
87		4.12	9.21			
	10			9.11	91.1	705.35
97		4.03	9.01			
	10			8.75	87.5	792.85
107		3.79	8.48			
	10			7.80	78	870.85

117		3.18	7.11			
	10			6.58	65.8	936.65
127		2.7	6.04			
	10			5.47	54.7	991.35
137		2.19	4.90			
	10			5.27	52.7	1044.05
147		2.52	5.64			
	10			4.75	47.9	1091.95
157		1.76	3.94			
	10			2.71	27.1	1119.05
167		0.66	1.48			

u/s = 1,119
d/s = 1,119
both = 2,238m²

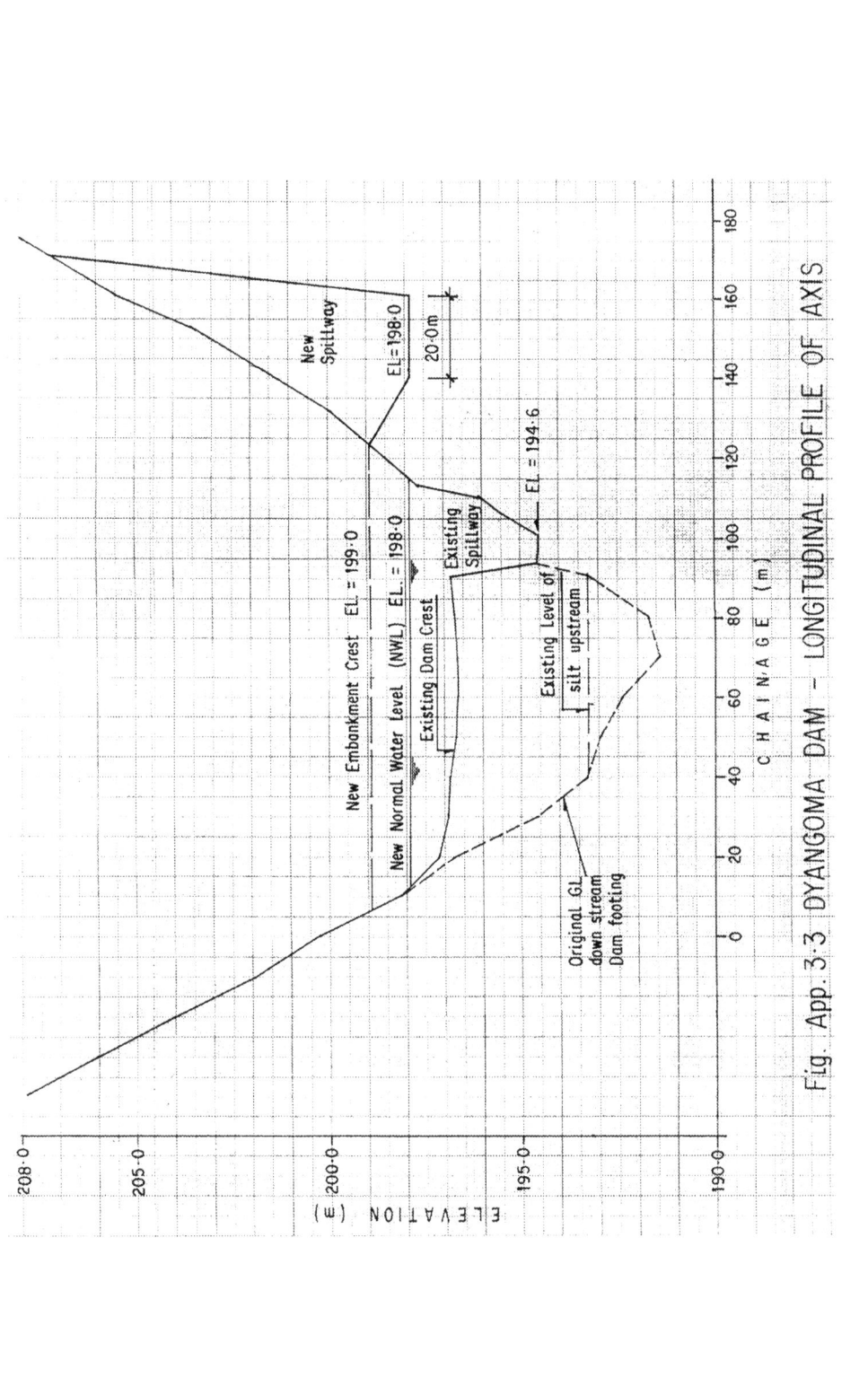

Fig. App. 3.3 DYANGOMA DAM - LONGITUDINAL PROFILE OF AXIS

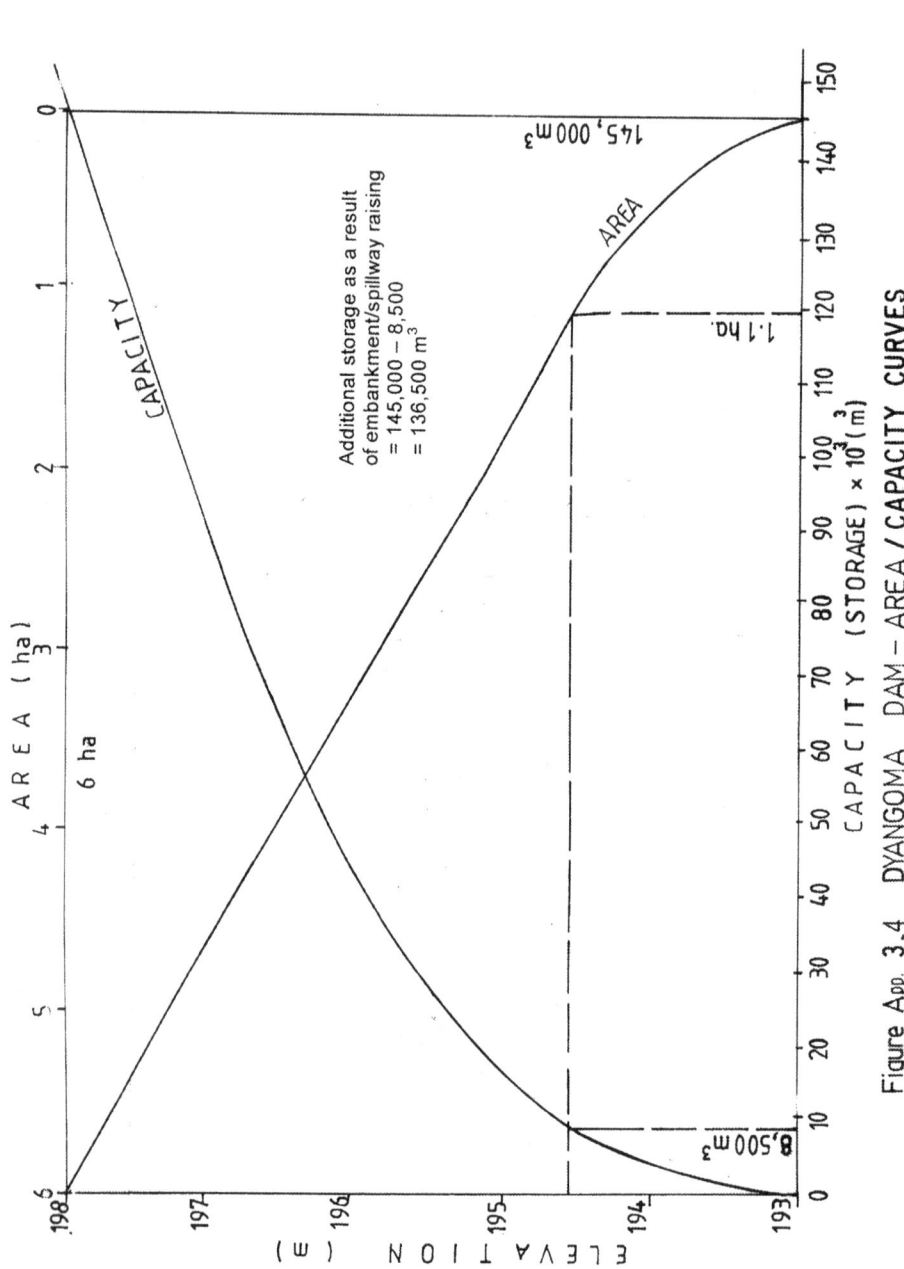

Figure App. 3.4 DYANGOMA DAM – AREA / CAPACITY CURVES

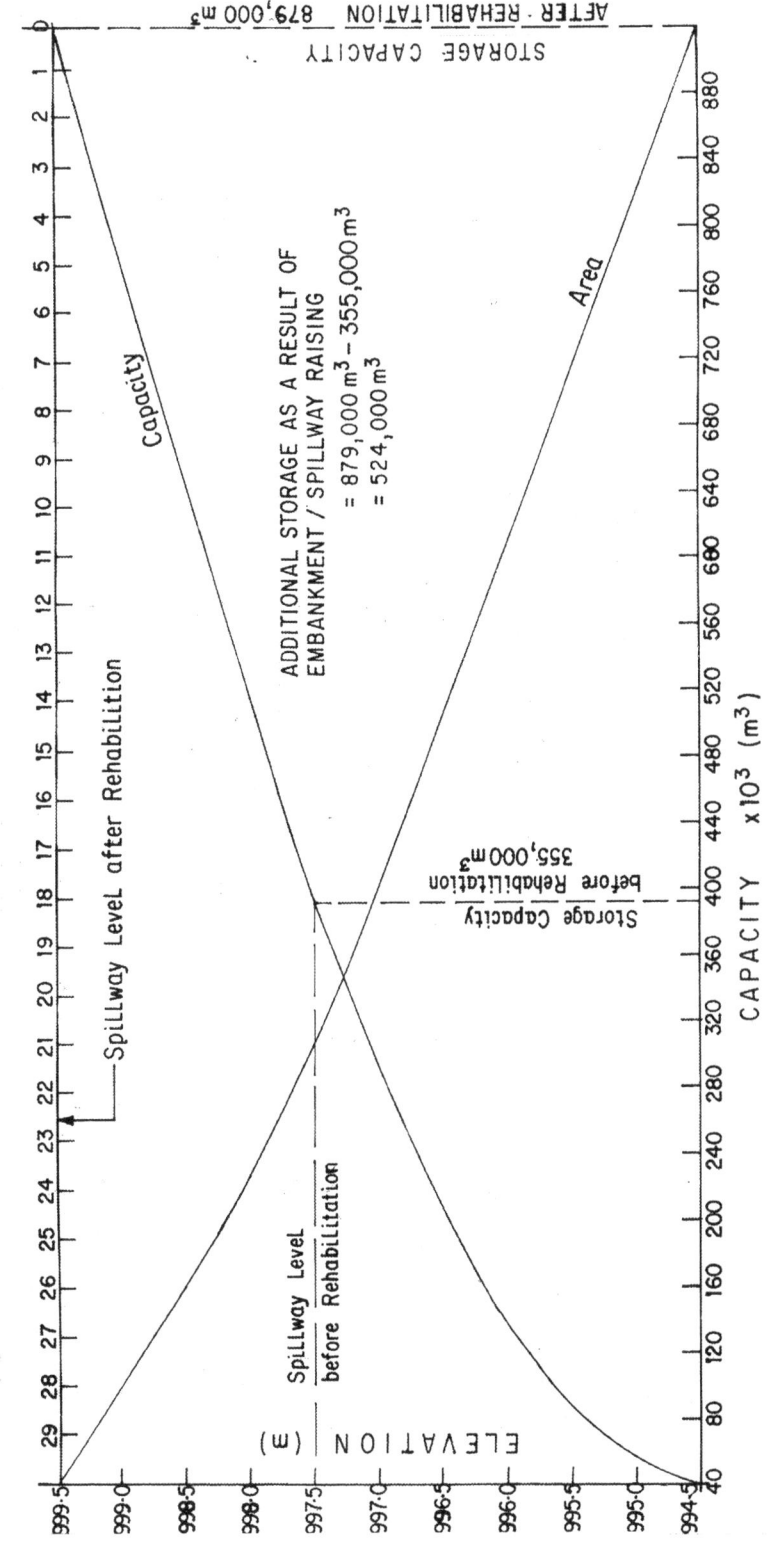

Figure- App. 3.5 **RWAMAKARA DAM REHABILITATION – AREA / CAPACITY CURVES**

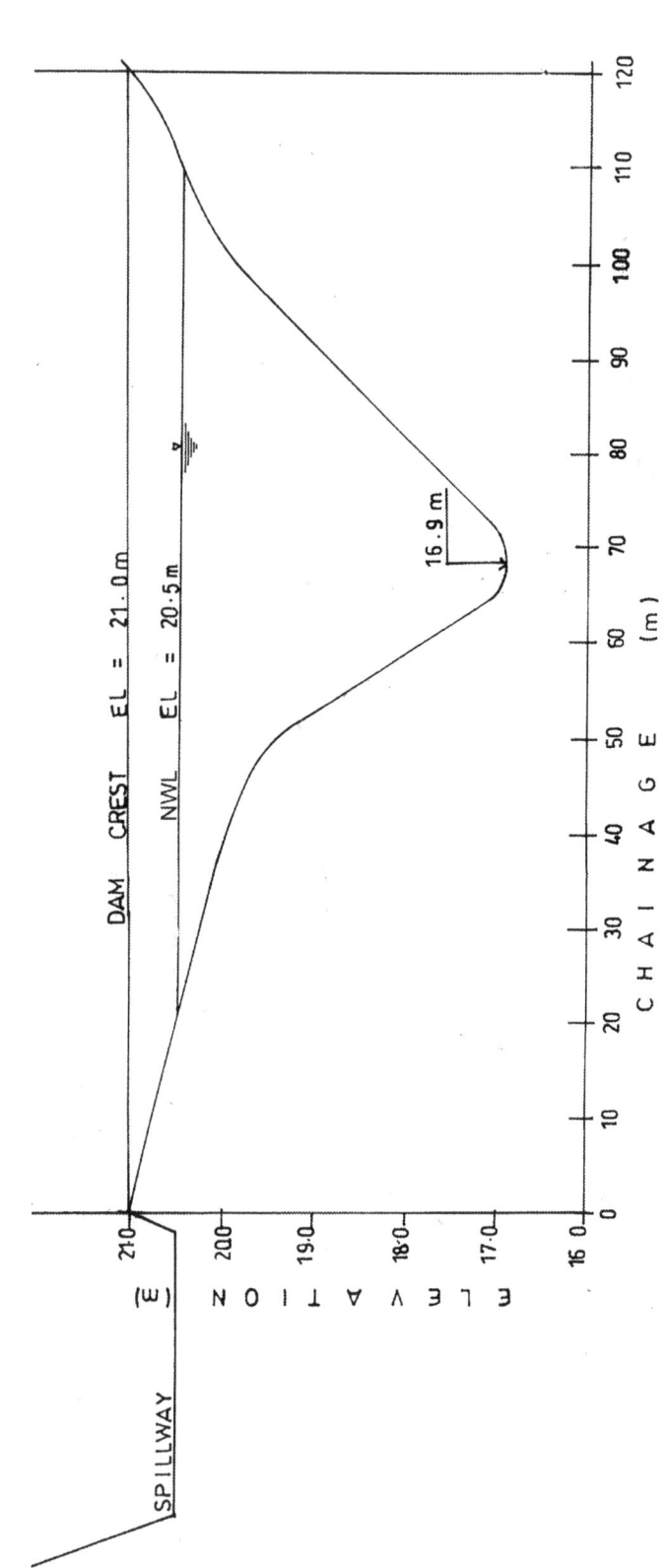

Figure-App. 3.6 RYERA DAM LONGITUDINAL PROFILE OF DAM WALL

Appendix IV

KENWA DAM CALCULATIONS

1. Calculation of Embankment Quantities for Upstream & Downstream Slopes of 2:1
2. Spillway Longitudinal Profile and Calculation of its Quantity

KENWA DAM, MBARARA DISTRICT – CALCULATION OF EMBANKMENT QUANTITIES: DOWNSTREAM SLOPE (D/S) = 2:1; UPSTREAM SLOPE (U/S) = 2:1

1	2	3	4	5	6	7	8	9	10	11	12	13
CHAIN-AGE (M)	INTERVAL L_1-L_0 (M)	ELEVATION (M)	HEIGHT $H = EL_{MAX} - EL_0$ (M)	CORE AREA $4(H-1)$ (M²)	EMBANK-MENT AREA $H(4+2H)$ (M²)	SHELL AREA $COL_6 - COL_5$ (M²)	AVERAGE CORE AREA (M²)	AVERAGE SHELL AREA (M²)	CORE VOL $COL_8 \times COL_2$ (M³)	CUMU-LATIVE CORE VOL (M³)	SHELL VOL $COL_9 \times COL_2$ (M³)	CUMU-LATIVE SHELL VOL (M³)
34		50		0	0	0	-					
40	6	47	3	8.0	30	22	40	11	24	24	66	66
50	10	46.9	3.1	8.4	31.62	23.22	4.2	22.61	42	66	226.1	292.1
60	10	48.4	1.6	2.4	11.52	9.12	5.4	16.17	54	120	161.7	453.8
70	10	46.9	3.1	8.4	31.62	23.22	5.4	16.17	54	174	161.7	615.5
80	10	45.8	4.2	12.8	52.08	39.28	10.6	31.25	106	280	312.5	928
90	10	44.7	5.3	17.2	77.38	60.18	15	49.73	150	430	497.3	1425.3
100	10	44.1	5.9	19.2	93.22	73.62	18.4	66.9	184	614	669	2094.3
110	10	44.0	6.0	20.0	96	76	19.8	24.81	198	812	748.1	2842.4
120	10	44.0	6.0	20.0	96	76	20	76	200	1012	760	2842.4
130	10	43.8	6.2	20.8	101.68	80.88	20.4	78.44	204	1216	784.4	3602.4
	10						20.5	79.65	205	1528	796.5	4386.8

1 CHAIN-AGE (M)	2 INTERVAL L_1-L_0 (M)	3 ELEVATION (M)	4 HEIGHT $H = El_{MAX} - El_0$ (M)	5 CORE AREA $4(H-1)$ (M^2)	6 EMBANKMENT AREA $H(4+2H)$ (M^2)	7 SHELL AREA $COL_6 - COL_5$ (M^2)	8 AVERAGE CORE AREA (M^2)	9 AVERAGE SHELL AREA (M^2)	10 CORE VOL $COL_8 \times COL_2$ (M^3)	11 CUMULATIVE CORE VOL (M^3)	12 SHELL VOL $COL_9 \times COL_2$ (M^3)	13 CUMULATIVE SHELL VOL (M^3)
140		43.9	6.1	20.4	98.82	78.42						
	10						19.8	74.85	198	1719	748.5	5183.3
150		44.2	5.8	19.2	90.48	71.28						
	10						18.8	69	188	1907	690	5931.8
160		44.4	5.6	18.4	85.12	66.72						
	10						15.4	52.17	154	2161	521.7	7143.5
170		45.9	4.1	12.4	50.02	37.62						
	10						10.2	29.81	102	2263	298.1	7441.6
180		47.0	3.0	8.0	30	22						
	10						4.4	14.44	44	2307	144.4	7586
190		48.8	1.2	0.8	7.68	6.88						
	8						0.4	3.44	3.2	2310	27.52	7613.5
198	-	50	0	-								
									Add core foundation	443		
									Total	2753		7614

KENWA DAM – CALCULATION OF EMBANKMENT SLOPE AREA: D/S = 2:1; U/S = 2:1

1	2	3	4	5	6	7	8
CHAINAGE (M)	INTERVAL L_1-L_0 (M)	ELEVATION (M)	HEIGHT $EL_{MAX} - EL_0$ (M)	SLOPE LENGTH $\sqrt{h^2+(2h)^2}$ (M)	AVERAGE SLOPE LENGTH (M)	SLOPE AREA (M^2) $COL_6 \times COL_2$	CULULATIVE SLOPE AREA (M^2)
31.5	8.5	51	0	0			
40	10	47	4.0	7.96	3.98	33.83	33.83
50	10	46.9	4.1	8.16	8.06	80.6	114.43
60	10	48.4	2.6	5.17	6.67	66.7	181.13
70	10	46.9	4.1	8.16	6.67	66.7	247.83
80	10	45.8	5.2	10.35	9.26	92.6	340.43
90	10	44.7	6.3	12.53	11.44	114.4	454.83
100	10	44.1	6.9	13.73	13.13	131.3	586.13
110	10	44.0	7.0	13.93	13.83	138.3	724.43
120	10	44.0	7.0	13.93	13.93	139.3	863.73
130	10	43.8	7.2	14.33	14.13	141.3	1005.05
140		43.9	7.1	14.13	14.23	142.3	1147.33

1	2	3	4	5	6	7	8
CHAINAGE (M)	INTERVAL L_1-L_0 (M)	ELEVATION (M)	HEIGHT $EL_{MAX} - EL_0$ (M)	SLOPE LENGTH $\sqrt{h^2+(2h)^2}$ (M)	AVERAGE SLOPE LENGTH (M)	SLOPE AREA (M^2) $COL_6 \times COL_2$	CULULATIVE SLOPE AREA (M^2)
150	10	44.2	6.8	13.53	13.83	138.3	1285.63
160	10	44.4	6.6	13.13	13.33	133.3	1418.93
170	10	45.9	5.1	10.15	11.64	116.4	1535.33
172	2	47.2	3.8	7.56	8.86	17.72	1553.05
180	8	47.0	4.0	7.96	7.76	60.08	1613.13
190	6.5	46.8	4.2	8.36	8.16	53.04	1666.17
200	3.5	48.8	2.2	4.38	6.37	22.295	1688.465
205	10	50.3	0.7	1.26	2.82	28.2	1716.665
210	10	51.0	0	0	0.63	6.3	1722.965
	5				–	–	1723
				U/S =	1723		
				D/S =	1723		
				Both	3446		
				S/W control	200		
				Filter belt reduce to	10,000		
					13,646 – save 288		

Kenwa Dam – Spillway Excavation Quantities

Crest: elevation 50 m

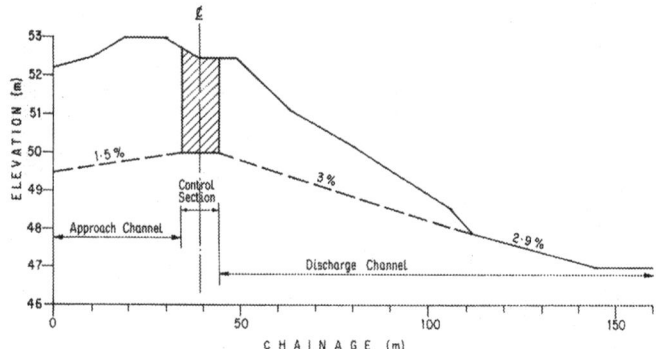

Figure: Appendix 4: Kenwa Dam, Spillway Longitudinal Profile

Approach
1. H = 2.87, top width = 28.61
   A = 972.74; B = 680; $\sqrt{AB}$ = 813.13
   V = 2359.18m³

Control
   H = 2.6; top width = 27.8
   A = 234.5; B = 200; $\sqrt{AB}$ = 235.80
   V = 618.62 m³

Discharge
1. H = 2.13, top width = 26.4
   A = 501.60; B = 380; $\sqrt{AB}$ = 436.59
   V = 935.91 m³

2. H = 1.2, top width = 23.6
   A = 1015.8; B = 860; $\sqrt{AB}$ = 934.2
   V = 1123.6 m³

3. H = 0.25, top width = 20.75
   A = 207.5; B = 200; $\sqrt{AB}$ = 203.72
   V = 50.93

4. H = 0.275, top width = 20.825
   A = 312.375; B = 300; $\sqrt{AB}$ = 306.13
   V = 84.20m³

Discharge Total = 2194.65 m³;
S/W Total = 5172.4 m³

Appendix V
SAMPLE SPECIFICATIONS AND BILLS OF QUANTITIES

A. Sample Specifications

1. INTRODUCTION

This is an example of the specifications for a dam project involving the construction of new dams and valley tanks and the rehabilitation of some, primarily for livestock watering; the specifications are applicable to both rehabilitation and the construction of new dams and valley tanks and will be applied as appropriate. The contractor is expected to adopt the proper construction methods which include (but not limited to):

a. adequate foundation preparation and treatment to the satisfaction of the engineer, prior to the placement of any core or fill material
b. proper placement of successive layers of core and fill materials on the embankment with the necessary watering and compaction, and
c. proper adherence to these specifications and the site-specific drawings, realising that each structure is designed to suit its specific site geology; any deviation shall be with the express authorisation of the engineer (or project manager).

2. DEFINITION OF WORKS

2.1 Rehabilitation
Rehabilitation of dams is generally made up of reservoir desilting/expansion, embankment and spillway repair, raising of embankment and spillway heights, installation/repair of drainage and water delivery devices and cattle troughs. The earth works are aimed at achieving an average reservoir depth of at least 3 m to reduce evapotranspiration to its barest minimum.

2.2 Valley Tanks
Rehabilitation is aimed at desilting the tank as well as deepening and widening it to the specifications given in the drawings.

2.3 Construction of new dams
This consists of building the embankment and spillway, as well as the drainage and water delivery system, from scratch. Wherever feasible, borrowing should be carried out within the reservoir area to maximise its storage potential, though without impairing its gravity delivery potential.

2.4 Construction of new valley tanks

This consists essentially of excavation works and the building of the appurtenant embankments and others relating to protection and water delivery devices.

3. THE EMBANKMENT

3.1 The Foundation

The essential requirements of foundation treatment are that it provides:

(i) a suitable support for the embankment under all conditions of saturation and loading, and

(ii) sufficient resistance to seepage which may endanger the stability of the dam as well as lead to considerable loss of water.

The following specifications are aimed at achieving just that.

3.2 Removal of Topsoil

The whole of the dam and reservoir areas (or the periphery of the NWL, where the total area to clear would exceed 10 hectares) shall be stripped of trees, shrubs and grass and any existing bonds, embankments and tanks/depressions in the reservoir, and levelled out to prevent detention storage that may delay reservoir filling and encourage run-off losses. The topsoil shall also be removed to firm solid material which will form the dam foundation and from which level the reference point for the valley tank depth shall be taken. The topsoil shall be stockpiled outside the dam/tank area for subsequent use for topsoiling; any unused material shall be carted to areas where they cannot be washed back into the reservoir. The reservoir area should be cleared early in the construction programme to encourage the regrowth of grass, which would help minimise soil movement in the course of reservoir filling on completion; this action would reduce the amount of silt getting into the reservoir.

3.3 The Clay Core or Water Barrier Portion

The foundation of the core shall be excavated into the impervious material as indicated in the drawing and approved by the engineer; this makes the depth variable. It shall however have a minimum bottom width of 4–5 m, depending on the width of the construction equipment.

3.4 Embankment Breaching (for rehabilitation only).

At appropriate locations, the embankment shall be breached to drain the reservoir and create a dry working environment and to install drainage/outlet pipes. The walls of the breached portion shall be terraced to enhance proper interfacing between the old and the new wall materials.

3.5 Embankment Construction (Core and Shell)

The core foundation shall be filled with impervious material (clay) and compacted to approximately 95% maximum dry density (MDD). Before placing the first layer of clay into the trench, the foundation shall be moistened and compacted with a tamping roller to enhance proper bonding between the foundation and the core. Any encountered rock foundation should be similarly moistened but no standing water should be allowed. The placement and compaction shall be as specified for fill materials below. The height of the core shall be 0.6 m above the normal water level. The remaining 0.4 m to attain a freeboard of 1.0 m above the spillway level shall be filled with the same materials as for the shell, in the same manner described below.

3.6 The Dam Wall (or Shell)

Well-drained gravelly materials devoid of organic matter (see Table 6.3) shall be laid around the core (i.e. upstream and downstream) in successive layers that are well blended, levelled and moistened uniformly to the optimum level required for maximum compaction. Compaction shall be by the passage twelve times of Tamping (or sheep-foot) non-vibratory rollers, towed by crawler-type or tyre-type tractors, at a speed not exceeding 8 km per hour) or self-propelled at the same speed limit, and shall be carried out in a series of continuous operations over the full width of the zone concerned. The compaction shall achieve the specified density of 95% maximum dry density (MDD), and the post-compaction thickness of each layer shall not exceed 150 mm.

Where smooth roller is used, the surface of the just compacted layer shall be scarified to a minimum depth of 50 mm to ensure proper interfacing between successive layers.

3.7 Embankment Slope and Crest Protection

The upstream and downstream slopes shall be trimmed and battered to 1:2[17] to enhance stability and protection against sudden draw-down of the reservoir. Ideally, the upstream slope should be rip rapped and the downstream slope grassed to provide protection against wave, wind and storm water erosion. But where funding is limited, especially where wave action is limited as where reservoir fetch (total length perpendicular to the embankment at NWL) is less than 1 km, both slopes could be grassed; the grass should however be well established before reservoir filling commences. The dam crest shall be similarly protected with a 150 mm lateritic material as final crest surface; crest surface protection will be increased to 300 mm of lateritic material where the top of the embankment will serve as a motorable road.

3.8 Scour/Drainage and Outlet Works

A 150–200 mm Drainage/Outlet pipe is to be embedded in the embankment for emptying the reservoir for safety and maintenance purposes (the scour/drainage function), and supply to cattle troughs downstream. The pipe shall be installed at the bed level of the reservoir with an aluminium strainer embedded in a 0.5 m high concrete sump to serve as sand filter. The grove for the pipe shall be cut manually from a firm ground, preferably below any fill area across the dam, just wide enough to house the pipe and cleaned of all loose materials. To prevent percolation along the pipe, concrete collars (or

---

[17] This varies, depending on design. See Table 6.3 also.

puddle flanges) should be made around the pipe in at least three locations along the pipe length that will eventually come under the embankment. Water-tightness at the joints will be achieved by means of a solvent cement 'which remains unaffected by almost all fluids for which PVC pipes are suitable' (Manufacturer's Note). At the exit of the 150 mm pipe downstream, a (preferably two) 150 or 200 mm gate valve and another one of 50 mm shall be fitted for drainage and supply to cattle trough respectively, using a 150/50 mm reducer tee as appropriate. The pipeline shall be tested for leakage and necessary corrections made before back-filling.

3.9 Back-filling

The back-fill around these pipes and valves shall be carefully placed and compacted manually/mechanically to the same density as that of the surrounding embankment, using a manual vibratory roller/compactor after the pipes have been carefully covered with the same materials as the surrounding embankment, i.e. clay in core and well-drained material in shell.

3.10 Stilling Basins

Stilling basins filled with rubble shall be provided at the exit of the drain to prevent erosion (see Figure 6.4C).

4   THE SPILLWAY AND FREEBOARD

The spillway shall be cut *in situ* to allow for a freeboard of about 1,000 mm, including an allowance for settlement of the fill material from the top of the dam to reduce the chances of over-topping. It shall be graded to site-specific approach and discharge slopes and led back to the original watercourse downstream of the dam, clear of the toe of the dam wall; the discharge end shall be protected against erosion with stone rubble. The side slopes of the spillway shall be graded to a slope of 1:1.5.

5.   THE ENVIRONMENTAL PROTECTION WORKS

5.1 Fencing

Fencing of the embankment and reservoir areas with doublestrand barbed wire (minimum gauge 16) on seasoned and treated timber to be supplemented with hedges later by the beneficiaries, using drought-resistant shrubs common in the area.

5.2 Grassing

Grassing all topsoiled excavated heaps and providing a grassed strip of about 30 m upstream of the reservoir, using paspallum or bahama. These grasses establish quickly and are drought resistant. The minimum thickness of compacted topsoil (light compaction) shall be 50 mm and Grassing shall be at 200 mm centres, watered and fertilised until it is properly established. It should be well established for the grass strip at the upstream end of the reservoir and embankment upstream slope before reservoir filling commences.

6. AS-BUILT DRAWINGS

The contractor shall, under the supervision of the engineer, prepare and deliver to the employer at the completion and handing over of the works, the original and three printed copies of the as-built drawings of all aspects of the works. The topographical drawing of the dam and reservoir areas shall indicate, among others, the elevations of the reservoir floor prior to impoundment.

7. SPECIFICATIONS

The specifications could be more or less than presented here, depending on the level of supervision envisaged during construction.

B. Bills of Quantities

1. This is an example of the notes that usually precede bills of quantities.
2. Objectives.
   The objectives of this Bill of Quantities are:
   a. to provide sufficient information on the quantities of works to be performed to enable bids to be prepared efficiently and accurately, and
   b. when a contract has been entered into, to provide a priced bill of quantities for use in the periodic valuation of the works.

In order to achieve these objectives, works should be itemised in the bill of quantities in sufficient detail to distinguish between different classes of works, or between works of the same nature carried out in different consideration of cost. Consistent with these requirements, the layout and content of the Bill of Quantities should be as simple and as brief as possible.

General Directions
1. This bill of quantities must be read in conjunction with the conditions of contract, agreement and the specifications and the contractor shall be deemed to have examined the drawings, specifications and agreement and to have acquainted himself with the detailed description of the works to be done and the way in which they are to be carried out, and specific requirements and standards of the finished works.
2. The quantities set down against the items in this bill of quantities are an estimate of the quantity of each kind of work included in the contract and are given for the convenience of forming a common basis for tenders. They are not to be taken as a guarantee that the quantities will be carried out or required or that they will not be exceeded.
3. The quantities shall therefore not be considered as representing the final measurements, it being the intention of the contract (except where otherwise specifically stated) that all works embraced therein shall be

measured upon completion by the project manager or his authorised representative and paid for at the prices and rates entered in the bill of quantities by the contractor.

4. The general description of materials given in the specifications is not necessarily repeated in the bill of quantities.

5. Each item in the bill of quantities is to be priced or if any items are left unpriced the contractor is to indicate that the value of the work described thereunder is elsewhere allowed for. No tender will be considered complete unless and until these requirements have been fulfilled. The prices and rates inserted under the several items include all costs and expenses which may be required in and for the construction of the work described under the several items including all costs and expenses which may be required in and for the construction of the work described together with all general risks, liabilities and obligations set forth or implied in these documents.

6. The bill of quantities is prepared in accordance with the current standard method of measurement of civil engineering quantities generally referred to in clause 57 of the conditions of contract (FIDIC) unless otherwise stated.

7. Earthworks quantities for filling, and clay-core are measured to compacted volumes within the required profile and the contractor must allow in his prices for the difference in volume during excavation or transport.

8. Earthwork quantities for excavation for materials which are not subsequently to be used as fill or clay-core, e.g. desilting and embankment breaching, are to be measured *in situ* before excavation.

9. Haulage rates will be assessed on the basis of the use of the nearest acceptable road taking materials to the section of the work under construction.

10. Trench and foundation excavation shall be measured to the net plan area required for the construction of the particular works. Measurement shall normally be based on excavation with vertical sides unless otherwise directed by the engineer.

11. The rate for all excavations shall allow for all necessary support or timbering and the removal thereof as necessary. In addition the appropriate rates shall allow for the exclusion of water from the excavation as necessary.

12. Shuttering shall be measured to the net area of surface of finished work to be supported during concreting.

13. Reinforcement shall be measured from the sizes and net lengths of the bars shown on the drawings or elsewhere and no allowance shall be made for maps, waste, spacer bars, etc.

14. Concrete shall be measured to the net size of the finished work and no deductions shall be made for the volume of reinforcement or for chambers not exceeding 150 mm by 150 mm for holes, openings or other voids less than 0.03 m for additions made for fillets not exceeding 150 mm x 150 mm.
15. Prices for structural steelwork shall allow for the provision, transport and erection on site, all as specified or as shown on the drawings and to the satisfaction of the engineer. All necessary staging, plant, equipment, nuts, bolts, washers and any other materials required for the tendered rates for steelwork are included; no separate payment shall be made on these accounts.
16. Any variation as ordered from the billed quantities will in no way void or abate the conditions or provisions of the contract or relieve the contractor from the responsibility of constructing the work to the required standard and carrying out the maintenance specified. Any additional labour, transport or material required to make good faulty work or used in repair or maintenance will be at the expense of the contractor and will not be taken into account when determining the sum to be paid for each item or section of the works.
17. Payment for the provision, transport and removal of any item of plant engaged upon the works shall be included in the rates entered in the bills of quantities.
18. If during the execution of the works, any item of plant should in the opinion of the project manager prove to be unsuitable or out of repair so as to fail to perform the service required in the execution of the work, the contractor shall replace such plant with other suitable plant at his own cost and until the replacement has been satisfactorily made, the project manager or engineer may withhold from subsequent payments a sum representing such portion of the monies as the engineer shall certify to be reasonable.
19. The amount allowed under provisional sums and/or contingencies shall be expended in whole or in part or not at all as the project manager may direct. Where work to which a provisional sum relates has been ordered by the engineer and has been executed by the contractor, the work as executed will be valued in accordance with the conditions of contract.
20. The contractor shall be required to submit to the employer on completion and handing over, three sets of as-built drawings for all aspects of the works.

LAKE MBURO RESETTLEMENT DEVELOPMENT PROGRAMME
Bills Of Quantities For Dams and Valley Tanks in Kanyaryeru
SITE: Kanyaryeru MOW WORKS: Construction of new valley tank and related works

NO	ITEM	UNIT	QUANTITY	RATE (SH. '000)	AMOUNT (SH. '000)
	BILL NO 1: EARTHWORKS				
1.01	Clear trees and shrubs in tank & ancillary work areas (fenced area & area bounded by broken lines downstream)	ha	2.9		
1.02	Excavate material to create underground reservoir between elevations 51.5 and 48.0 m (3.5 m deep). Use the spoil to form banks around tank as per drawing; level and compact.	m^3	5,805		
1.03	Cut inlet channels as per drawing	m^3	20		
1.04	Excavate Silt trap and Stilling Basins	m^3	22		
1.05	Create contour bund – 1.5 m high along elevation 52.0 m to divert surplus flow (Spillway)	m^3	4,000		
1.06	Trim to design slopes on valley tank, silt trap and channels	m^2	1,500		
1.07	Supply and line up silt trap with in situ concrete 1:3:6	m^3	22		
1.08	a) Provide & lay 50 mm depth top soil, compact lightly and grass on spoil banks as protective cover against erosion	m^2	1,500		
	b) Ditto for spillway	m^2	150		
	TOTAL BILL NO 1				

LAKE MBURO RESETTLEMENT DEVELOPMENT PROGRAMME
Bills Of Quantities For Dams And Valley Tanks In Kanyaryeru
SITE: Kanyaryeru MOW WORKS: Construction of new valley tank and related works

NO	ITEM	UNIT	QUANTITY	RATE (SH. '000)	AMOUNT (SH. '000)
	BILL NO 2: FILTRATION WORKS				
2.01	Excavate filter chamber on end of valley tank and continue for filter channel and well 1000 mm internal diameter & 4,000 mm deep	m^3	400		
2.02	Provide, grade and lay filter materials: a. Hard core b. Graded aggregates 6–25 mm	m^3 m^3	80 40		
2.03	Provide, install and back fill reinforced concrete rings for the well in 2.01 above:	No	5		
2.04	Provide, cast and lay R.C. cover for the well – 200 mm x 1500 mm diameter with 2 No 40 mm diameter openings.	No	1		
	TOTAL BILL NO 2				

LAKE MBURO RESETTLEMENT DEVELOPMENT PROGRAMME
Bills Of Quantities For Dams And Valley Tanks In Kanyaryeru
SITE: Kanyaryeru MOW WORKS: Construction of new valley tank and related works

NO	ITEM	UNIT	QUANTITY	RATE (SH. '000)	AMOUNT (SH. '000)
	BILL NO 3: PLUMBING, OVERHEAD TANK AND TROUGHS				
3.01	Provide, lay and join 150 mm diameter PVC class D pipe for gravity supply	m	45		
3.02	Supply and fix Aluminium strainer for 3.01	No	1		
3.03	Provide & fix Reducer Tee 150/38 mm	No	1		
3.04	Provide, lay & join 38 mm diameter G. I. pipe with provision to connect to pump from well to o/head tank.	m	6		

NO	ITEM	UNIT	QUANTITY	RATE (SH. '000)	AMOUNT (SH. '000)
3.05	a. Ditto from tank to troughs and stand taps. b. Provide fix stand taps with Fittings	m No	14 4		
3.06	Provide and align 10 m³ reinforced concrete/block o/head tank – 2.5 m diameter & 2 m high as per drawing	No	1		
3.07	a. Provide 4 HP motorised petrol engine pump (Yamaha or Honda) with 7 m suction & 25 m delivery & discharge of minimum 10 l/sec. with appropriate connection couplings. b. Supply & install NIRA AF-85 with 63 mm diameter cylinder, 15 m suction & 33 l/sec. Discharge	No No	1 1		

NO	ITEM	UNIT	QUANTITY	RATE (SH. '000)	AMOUNT (SH. '000)
3.08	Construct 2 No Watering Troughs	No	2		
	- Block work	m^2	11.3		
	- Concrete 1:3:6	m^3	9		
	Hard stand				
	- Hard core	m^3	19		
	- In situ concrete 1:2:4 over hard core	m^3	4		
	Plastering & Rendering				
	- External rough finish	m^2	11.3		
	- Internal fine finish	m^2	10		
	- 2 float valves, 2 gate valves, plus fittings	Nos	2,2 etc		
	- Concrete protection for float valve	No	1		
	TOTAL BILL NO 3				

LAKE MBURO RESETTLEMENT DEVELOPMENT PROGRAMME

Bills Of Quantities For Dams And Valley Tanks In Kanyaryeru

SITE: Kanyaryeru MOW WORKS: Construction of new valley tank and related works

NO	ITEM	UNIT	QUANTITY	RATE (SH. '000)	AMOUNT (SH. '000)
4.0	BILL NO 4: FENCING				
1	Provide 2 strand 16 gauge barbed wire, to be drawn taught, 6 lines on 150 mm diameter solignum treated hard wood poles (Eucalyptus) at m c/s and 1 intermediate batten strainers at corners. Poles at corners, centre & gates and every 10th position should be erected in concrete 1:3:6.	m	710		
2	Provide well framed hard wood gate in fence to allow for inspection and maintenance.	No	1		
	TOTAL BILL NO 4				

Kanyaryeru Mow – Bill Summary For: Construction Of A New Valley Tank & Related Works

BILL NO	ITEM	AMOUNT (SH.)
1.	Earthworks	
2.	Filtration works	
3.	Plumbing, overhead tank & Troughs	
4.	Fencing	
SUBTOTAL CONTINGENCY (10%)	- -	
GRAND TOTAL		

NAME OF FIRM ....

SIGNATURE OF REPRESENTATIVE ....

DESIGNATION OF SIGNATORY ....

Appendix VI
ABRIDGED BIDDING (TENDER) DOCUMENT, BASED ON FIDIC FORMAT

TITLE PAGE

Consultant   July 1995

CONTRACT DOCUMENTS

VOLUME I:

TENDER DOCUMENTS

CONDITIONS OF TENDER, FORM OF TENDER, TENDER BOND,
APPENDIX TO TENDER,
SCHEDULE OF SUPPLEMENTARY
INFORMATION, AGREEMENT, PERFORMANCE BOND
and
BILLS OF QUANTITIES

TABLE OF CONTENTS

1. Background Information
2. Invitation for Bids
3. Description of the works
4. Conditions of Tender- General
5. Tender Opening
6. Process to be confidential
7. Classification of Tenders
8. Determination of Responsiveness
9. Correction of Errors
10. Conversion to single currency
11. Evaluation and Comparison of Tenders
12. Form of Tender
13. Appendix to Tender
14. Form of Tender Bond
15. Agreement
16. Form of Performance Bond
17. Schedule of Supplementary Information
18. Bills of Quantities- General Directions
    - Specific Conditions
    - Bills of Quantities for Dams

BACKGROUND INFORMATION

The ....

....
(hereinafter referred to as the Employer) is implementing the
....
Project financed by ....
(e.g. IFAD or IDA) and/or the Employer .... intends to apply part of the proceeds of these funds to implement the .... (name of project, its location & intended benefits)

Invitation for Bids
The Employer hereby invites bids for the........................... (nature of project e.g. Construction/Rehabilitation of Dams as described below. Bidding is open to all ....
(either Pre-selected or eligible tenderers)

Bidding documents may be purchased at.............................

for a non-refundable fee of .... per set between .... and .... on working days from ....

The purchase of bid documents closes on .... Interested bidders may obtain further information at the same address

DESCRIPTION OF THE WORKS

(Example):
The works to be carried out consist of the rehabilitation of

.... dams and the construction .... of new ones in
.... and .... Districts

CONDITIONS OF TENDER

1. Tenders shall be delivered to the office of ………………………………………………………… not later than 12.00 noon on the ……………… day ………………… Tenders received late shall be returned unopened to the tenderers. They shall be enclosed in a plain sealed envelope, endorsed, …………………………… and sent by hand and a receipt obtained or by registered post. The name of the tenderers must not appear on the envelope. The employer does not accept responsibility for whatever happens to unsealed tenders.
2. Every Tender shall be made on the form attached hereto without any alteration therein and duly signed by a Principal of the Company. With every Tender, there shall be a complete and unaltered copy of:-

(a) The Conditions of Contract Volume I

(this Volume) made up of:
- Conditions of Tender
- Form of Tender
- Appendix to Tender
- Schedule of Supplementary Information
- Tender Bond
    - Agreement
- Performance Bond
- Bills of Quantities

(b) The Conditions of Contract Volume II made up of:
- Part I: Conditions of Contract for Works of Civil Engineering Construction – General Conditions.
- Part II: Conditions of Particular Application
- Part III: Specifications

For the avoidance of doubt, the following forms are to be completed and submitted at tendering for the whole works or part thereof.

Form of Tender
Form of Tender Bond
Appendix to Form of Tender
Schedule of Supplementary Information
Bills of Quantities

Such Bills of Quantities shall have the appropriate columns filled up in ink by the person, firm or company making the Tender (hereinafter referred to as 'the Tenderer') with the rates and prices upon which the Tender is based.

3. The Agreement, Form of Performance Bond and Advance Payment Guarantee shall be completed only by the preferred Tenderer (hereinafter referred to as the contractor) who receives notification that his tender has been accepted and therefore requested to enter into such agreement.

4. The letter accompanying the Tender and other documents not supplied by the employer shall not form a part of the Tender.

5. If any alteration is made (other than filling in the blanks intended to be filled in the Tender), except by an official amendment, or if these instructions are not otherwise complied with, the Tender may be rejected.

6. A Tenderer who submits a Tender will be held to have satisfied himself independently as to the nature, extent and practicability of the works, and to have examined the site and its surrounds, the position relative thereto of existing works, buildings, services, means of access, the availability of materials, the nature of the soil or strata in or on which the works are to be constructed, and all other matters and circumstances in which the works are to be constructed, and all other matters and circumstances which may influence or affect his Tender. The costs and risks (if any) in obtaining all necessary information are those of the Tenderer.

7. The numbers, quantities and measurements set out in the Bills of Quantities are estimates only, and their accuracy or inaccuracy shall in no way affect the validity of the Tender or of any Contract based thereon. The total amount of each item set out in the Bills of Quantities at the rate of price inserted by the Tenderer shall be stated, but these figures are required solely for the purpose of facilitating a comparison of the various Tenders and shall not be deemed to be the actual sums which shall be paid to the contractor for the execution of the works. The sum to be paid to the contractor shall (subject to the provisions of the Conditions of Contract) be determined by measuring the work actually done in accordance with the contract and valuing it at the rates and prices inserted by the contractor in the Bills of Quantities.

8. The rates and prices inserted in the Bills of Quantities shall be the full inclusive rates and prices for the finished work described under the respective items and shall cover all labour, materials, transport, cartage, storage, temporary work, plant and overhead charges, watching, lighting and profit out of the conditions.

9. Should there be any discrepancy in or any doubt or obscurity as to the meaning of any of the contract Documents or as to anything to be done or

not to be done or as to any other matter or thing, the person tendering must set forth in writing such discrepancy, doubt or obscurity and submit the same as soon as possible, but not later than 14 days before the date fixed for delivery of Tenders, to the Headquarters of ……………, ………………, for elucidation. Similarly, any notice of withdrawal should reach the Employer not later than ……… (usually 14 days) before the closing date for the receipt of Tenders. Such notices shall be accompanied by unmutilated or otherwise defaced originals of all Tender Documents collected. No refund of Tender fees shall be made.

10. Documents will usually be issued to the contractors in duplicate, and both sets including drawings must be returned to the office of the ……………… by each Contractor by the closing date for Tenders; where both Documents are completed and the Bills priced, one shall be marked 'ORIGINAL'; this should be reflected on the conveying envelope as well. Following the acceptance of a Tender, all drawings and unpriced documents shall be returned to the ……………… with the exception of those held by the successful Contractor. Priced Bills of Quantities and Tender Bond will be returned to unsuccessful Tenderers.

11. ……………………………………… (the employer) do not bind themselves to accept the lowest or any Tender, nor to assign any reason for the rejection of any Tender.

Tender Opening

12. The Employer will open the Tenders, including submissions made pursuant to Clause 5 hereinbefore, in the presence of Tenderers' representatives who choose to attend, at the time and location indicated in the Letter of Invitation to Tender. The Tenderers' representatives who are present shall sign a register attesting to their attendance.

13. Tenders for which an acceptable notice of withdrawal has been submitted pursuant to Clause 9 hereinbefore, shall not be opened. The Employer will examine Tenders to determine whether they are complete, whether the requisite Tender Securities have been furnished, whether the Documents have been properly signed, and whether the Tenders are generally in order.

14. At the Tender opening, the Employer will announce the Tenderers' names, the Tender Sums, written notifications of Tender modifications and withdrawals, if any, the existence of the requisite Tender Security, and such other details as the Employer may consider appropriate.

15. The Employer shall prepare, for his own records, minutes of the Tender opening, including the information disclosed to those present in accordance with this clause.

Process to Be Confidential

16. After the public opening of Tenders, information relating to the examination, clarification, evaluation and comparison of Tenders and recommendations concerning the award of Contracts shall not be disclosed to Tenderers or other persons not officially concerned with such process until the award of the contracts to the successful Tenderer or Tenderers has been announced.

17. Any effort by a Tenderer to influence the Employer in the process of examination, clarification, evaluation and comparison of tenders will result in the disqualification of that Tenderer's Tender and the forfeiture of his Tender Security.

Clarification of Tenders

18. The Employer may ask Tenderers individually for clarification of their Tenders, including breakdowns of unit rates. The request for clarification and the response shall be in writing or by cable, but no change in the price or substance of the Tender shall be sought, offered or permitted except as required to confirm the correction of arithmetic errors discovered by the Employer during the evaluation of the Tenders in accordance with Clause 24 hereinafter.

Determination of Responsiveness

19. Prior to the detailed evaluation of Tenders, the Employer will determine whether each Tender is substantially responsive to the requirements of the Tender Documents.

    For the purpose of this Clause, a substantially responsive Tender is one which conforms to all the terms, conditions and specifications of the Tender Documents without material deviation or reservation. A material deviation or reservation is one which affects in any substantial way the scope, quality, or performance of the works, or which limits in any substantial way, inconsistent with the Tender Documents, the Employer's rights or the Tenderer's obligations under the contract, and the rectification of which deviation or reservation would affect unfairly the competitive position of other Tenderers presenting substantially responsive tenders.

20. If a Tender is not substantially responsive to the requirements of the Tender Documents, it will be rejected by the Employer, and may not subsequently be responsive by the Tenderer having corrected or withdrawn the non-conforming deviation or reservation.

Correction of Errors

21. Tenders determined to be substantially responsive will be checked by the Employer for any arithmetic errors in computation and summation. Errors will be corrected by the Employer as follows:
    (a) where there is a discrepancy between the amounts in figures and in words, the amount in words will govern; and
    (b) where there is a discrepancy between the unit rate and the total amount derived from the multiplication of the unit rate and the quantity, the unit rate as quoted will govern, unless in the opinion of the Employer there is an obvious gross misplacement of the decimal point in the unit rate, in which event the total amount as quoted will govern and the unit rate will be corrected.

22. The amount stated in the Tender Form will be adjusted by the Employer in accordance with the above procedure for the correction of errors and, with the concurrence of the Tenderer. If the Tenderer does not accept the corrected amount of Tender his Tender will be rejected and the Tender security will be forfeited.

Conversion to Single Currency

23. To facilitate evaluation and comparison of Tenders, the Employer will convert the amounts in various currencies in which the Tender Sum is payable (excluding Provisional Sums) to the United States Dollar at the selling rates established by the Bank of .... (country) 30 days prior to the Tender Date.

Evaluation and Comparison of Tenders

24. The Employer will evaluate and compare only Tenders determined to be substantially responsive to the requirements of the Tender Documents in accordance with Clause 19 above.

    In evaluating Tenders, the Employer will determine for each Tender the Evaluated Tender Sum by adjusting the Tender Sum as follows:
    (a) making any correction for errors pursuant to Clause 21 above;
    (b) excluding Provisional Sums and provision, if any, for Contingencies in the Summary Bill of Quantities,
    (c) converting the amount in (b) above to a single currency in accordance with Clause 23 above;
    (d) adding any benefits foregone for longer times of completion, assessed in accordance with this clause; and

(e) making an appropriate adjustment for any other acceptable quantifiable variations, deviations or alternative offers not reflected in the Tender Sum or in the above-mentioned other adjustments.

25. For the purpose of comparing different times of completion offered by respective Tenderers pursuant to Appendix to Tender, the benefits foregone by the Employer for each month of completion longer than the shortest time for completion offered by a responsive Tenderer shall be assessed probate at the rate of 12 per cent per annum of the respective Tenderer's Tender Sum (corrected for any arithmetic errors). The foregone benefits so assessed shall be discounted to present per annum and shall be added to the respective Tenderer's Tender Sum for comparison purposes only.

26. The Employer reserves the right to accept or reject any variation, deviation or alternative offer. Variations, deviations, alternative offers and other factors which are in excess of the requirements of the bidding documents or otherwise result in the accrual of unsolicited benefits to the Employer shall not be taken into account in bid evaluation.

27. Price adjustment provisions applying to the period of execution of the contract shall not be taken into account in bid evaluation.

28. The Employer reserves the right to negotiate with the Preferred Tenderer any of the unit rates or prices listed in the Bill of Quantities. If, in the opinion of the Employer, the unit rates or prices for some of the items in the Bill of Quantities are unreasonable or out of proportion, or the Tender as a whole is seriously unbalanced in relation to the Engineer's Estimate of the real cost of work to be performed under the contract; if the Tenderer fails, within a period of fourteen (14) days after having been notified in writing by the Employer to adjust the unit rates or prices concerned, the Employer may:

    1) require that the amount of the Performance Bond set forth in Appendix to Tender herein-below be increased at the expense of the Preferred Tenderer to a level sufficient to protect the Employer against financial loss in the event of subsequent default of the Preferred Tenderer under the contract, or
    2) reject the Tender.

29. On receipt of written notification from …………………… (employer) that his Tender has been accepted, the contractor shall without undue delay, whenever required by notice in writing, enter into and execute a Contract Agreement in the form hereto annexed prior to the expiration of his Tender Validity. The Contractor shall at his own expense obtain the Guarantee of a Bank approved by the Employer and duly empowered to enter into a Bond in the form annexed hereto as to be bound together with the contractor to the Government of ………………… through

………………… (employer) in the sum provided in the Tender for the due performance of the contract. He shall enter into a similar bond – Advance Payment Guarantee – if advance payment is granted.

30. Every direction or notice to be given to the contractor may be posted to the address given in the Tender and such posting shall be deemed good service of such notice, and the time mentioned in these Conditions for doing any act after direction of notice shall be reckoned from the time of such posting.

Dated this …………………day of ……… 20…
Signature ……………………………………………
in the capacity of …………………………… duly authorised to sign Tenders for and on behalf of ……………………………
……………………………………………………
……………………………………………………
(IN BLOCK LETTERS)

Address …………………………………………………..
……………………………………………………………..

WITNESS………………………………………………...
Address …………………………………………………..
……………………………………………………………..
Occupation ………………………………………………..

FORM OF TENDER BOND

Date.........................
Name of Bank..
Address of Bank..

.... agrees to make a Tender Bond to the Commissioner/Treasury Officer of Account,....................................... hereinafter called the Employer, on the following terms:

1. That the Tender of:
   Name of Tenderer..
   Address..
   ....
   is made in agreement with the Tender Documents for the
   ....
   (name of project).

2. In submitting the Tender the Tenderer agrees to furnish a Tender Bond through the above-named Bank in the sum of
   (2.5% of Tender sum)......................................
   ....

3. The above-named Bank hereto agrees to be guarantor to the Employer for the Tender Bond for the above sum.

4. If (Name of Tenderer) ....
   who has submitted the Tender does not abide by the Tender or any of the related conditions contained in the Tender Documents, the above-named Bank agrees to pay the sum of ....
   ....
   to the Employer within 7 (seven) days after receiving notification from the Employer for the default of the Tenderer.

5. This Tender Bond is effective from............................
   to .... (dates).
   The Bond shall be effective from the date of submission of the Tender until 120 (one hundred and twenty) days after the date fixed in the Letter of Invitation to Tender for receipt of the Tender.

IN WITNESS WHEREOF the authorised representative of the
....

Name of Bank:……………………………
Address:……………………………………
………………………………………………
hereunto signed and stamped.
Name of Witness: ……………………….
Address of Witness: …………………….

FORM OF TENDER

(NOTE: The appendix forms part of the Tender. Tenderers are required to fill in all the blank spaces in this Tender Form and Appendix).

TO: The (Employer) …………………………………...
………………………………………………………………….

Sir,
(project title/credit no)

Having examined the Drawings, Conditions of Contract, Specifications and Bills of Quantities for the construction of the above-named works, we offer to construct, complete and maintain the whole of the said works in conformity with the said Drawings, Conditions of Contract, Specifications and Bills of Quantities for the sum of:

(1) CONTRACT NO: ………………
Amount……………………………………………. (US$…………)

(2) CONTRACT NO: ………………
Amount……………………………………………. (US$…………)

(3) CONTRACT NO: ………………
Amount……………………………………………. (US$…………)

Or such other sum as may be ascertained in accordance with the said Conditions.

We undertake, if our Tender is accepted, to complete and deliver the whole of the works comprised in the contract within the time stated in the Appendix hereto.

If our Tender is accepted we will, when required, obtain the guarantee of a Bank (to be approved by the Employer) to be jointly and severally bound with us in a sum equal to 10 per cent of the above-named sum for the due performance of the contract under the terms of a Bond in the form annexed to the Conditions of Contract.

We agree to abide by this tender for the period of three (3) calendar months from the date fixed for receiving the same and it shall remain binding upon us and may be accepted at any time before the expiration of that period.

Unless and until a Formal Agreement is prepared and executed, this Tender, together with the Employer's written acceptance thereof, shall constitute a binding Contract between the Employer and ourselves.

We understand that the Employer is not bound to accept the lowest or any Tender he may receive.

APPENDIX VII

Bid Analysis and Evaluation

1. Sample Bid Analysis and Evaluation Format
2. Summary Tables from an Analysis and Evaluation Report.

Format for Bid Analysis and Evaluation for the Construction/Rehabilitation of Dams and Valley Tanks

Contract No: ....

Introduction

1. The objective of bid (tender) analysis and evaluation is to compare tenders (or bids) on the basis of their evaluated costs so that the bid with the lowest evaluated cost can be identified and decision makers advised accordingly. The lowest evaluated bid (not necessarily the lowest bid) is the preferred bid.
2. These guidelines are based on the provisions of the World Bank's *Standard Bidding Documents, Procurement of Works in smaller Contracts)*, June 1995 edition.

Bid Opening

3. Bids shall be opened publicly shortly after the closing of their receipt. This vital information has also been included in the bidding documents. Opening the bids publicly means opening them in the presence of bidders or their representatives, if they wish to attend. This will give them the necessary confidence that their bids would receive fair treatment during analysis and evaluation. The bidders should however sign their names in a register as a record of their presence.
4. Envelopes marked 'withdrawal' shall be opened and read out first. Thereafter, the chairman of the bid opening meeting should take the tenders one after the other in a random manner, open them and announce to the hearing of all present, the name of the bidder and the bid amount, and any other given details as required by the bidding document e.g. bid security. These should then be recorded in a table, a sample of which is given here as Table 1. The form should be duly signed and dated by all members of the committee and attached to the minutes of that meeting which in itself would be an attachment to the bid analysis and evaluation report.
5. After the opening of the bids and the announcement and recording in the table mentioned above, the rest of the bid analysis and evaluation should be done confidentially.

Bid Analysis and Determination of Responsiveness

6. The bids should thereafter be subjected to a thorough scrutiny to ensure that they comply with the requirements of the tender as contained in the bidding document; more specifically, that:
   i. they meet the eligibility requirements of the International funding Institutions (IFI), IDA in this case
   ii. they are properly signed
   iii. they are accompanied by the required bid security
   iv. there are no material (substantial) deviations from the requirements of the bidding documents.

   In short, to determine whether they are substantially responsive to the bidding document. Any bid that deviates from the bidding conditions generally referred to as being non-responsive should not be considered further.

7. Responsive bids are then checked for arithmetic errors; if errors are found, they should be corrected as follows:
   i. where there is a discrepancy between the amounts in figures and in words, the amounts in figures will govern
   ii where there is a discrepancy between the unit rate and the line item total resulting from multiplying the unit rate by the quantity, the unit rate as quoted will govern, unless, in the opinion of the employer, there is an obvious gross misplacement of the decimal point in the unit rate, in which case, the line item total as quoted will govern and the unit rate will be corrected.

8. The bid amount will be adjusted by the employer in accordance with the procedure stated above, and with the concurrence of the bidder, shall be considered as binding upon the bidder. If the bidder does not accept the corrected amount, the bid will be rejected and the bid security may be forfeited in accordance with sub-clause 16.6(b) of the bidding document.

9. if there are ambiguities in any bid that may make comparison difficult or impossible, such bidder could be asked for clarification in writing, but under no condition should any bidder be allowed to revise their bid once the bids have been opened.

10. It is usually helpful to consider each bidder's bill items with the engineer's estimate which forms a good basis for the comparison of bids. Tables 2 and 3 are samples of such tables of comparison.

11. Provisional and contingency amounts are usually left out of the bid amounts for the purpose of comparison.

12. Any offered or permitted variations/deviations and alternatives are analysed and credited to the relevant bidder(s).

13. Any discounts and/or price modifications are similarly adjusted for.
14. Since there is not likely to be a pre-qualification of tenderers in the case of small dams because of the small-scale nature and simplicity of the structures involved, a critical analysis of the qualification information in the bidding documents will help assess the capability of each tenderer in terms of personnel, equipment, financial resources, experiences in similar works etc. the minimum number and types of equipment recommended per construction site are given in the appendix to this format.
15. It is important to note that the validity of the bid security should be at least 30 days longer than the period of validity of the bid itself; this is to provide sufficient cover for the bid, in case it becomes necessary to extend the period of bid validity before the contract can be signed.
16. The schedule of construction proposed by each bidder should be examined and compared with the engineer's schedule in relation to the weather conditions of the project area, the equipment proposed, past experience on similar works and whether it is possible to achieve the target set by the contractor.

Bid Evaluation and Comparison

17. On the basis of the foregoing considerations, all substantially responsive bids will be compared. The bidder whose evaluated bid price(s) is/are lowest emerge(s) as the preferred bidder(s) and should accordingly be recommended to the Contract Awarding Authority (e.g., the Central Tender Board), either for further negotiation if there are outstanding issues to clear or for contract award.

    For example, in summarising the analysis report a possible situation could be:

    Preferred Bidder X

    Comments:
    i. although their bid price is the second lowest, considering the discount offered if an advance payment of 15% is granted, their bid price becomes the lowest
    ii. the foreign currency component requested is the lowest, being only 8% of the bid price
    iii. they have not imposed any reservations and/or conditions that could increase the bid price
    iv. they have thoroughly studied the conditions of bidding, acquainted themselves with site conditions and submitted a detailed method of construction.

Recommendation

18. Based on the foregoing analysis and appraisal, we recommend that the employer invites Bidder X for final negotiation to secure the best possible arrangement for constructing and rehabilitating all (or part thereof) batches of the Project.

Contract Award

19. As much as possible, the ensuing contract should be signed with the preferred bidder in accordance with the procedure outlined above soon after the decision to this effect has been taken and within the validity period of the tender. The following is the sequence of events that follows a successful bid analysis and evaluation exercise:
20. The employer notifies the successful bidder(s) by forwarding the letter(s) of acceptance and requests of them performance securities.
21. The successful bidders forward performance securities to the employer.
22. The employer notifies the unsuccessful bidders that their bids were not successful.
23. The employer signs the contract agreement and forwards it to the contractor to sign or they both sign at the same time and place; the latter option is generally preferred.

TENDERS FOR THE CONSTRUCTION /REHABILITATION OF DAMS

Table 1   Record of tenders as recorded at opening on _____

S/N	NAME OF BIDDER (OR BIDDER IDENTIFICATION	BID SUM (US$)	REMARKS[18]
1.			
2.			
3.			
4.			
5.			
6.			

Note: this table assumes that there are six (6) respondents; the table can be extended as necessary.

Signatures of witnesses: 1.   …………………………..
                                           2   …………………………..
                                           3   …………………………..
                                           4   …………………………..

See a real-life example in Table A.

---

[18] Under this column, record whether the tender contains all that is required, any additional information etc.,

TENDERS FOR THE CONSTRUCTION /REHABILITATION OF DAMS

TABLE 2   BID AMOUNTS BEFORE AND AFTER CORRECTION COMPARED WITH ENGINEER'S ESTIMATE

NO	BIDDER	BID SUM TENDERED (US$)	SUM AS CORRECTED (US$)	DEVIATION FROM ENGR'S (%)	DISCREPANCY (+/-) (US$)
1.					
2.					
3.					
4.					
5.					
6.					

See a variant of this table in Table C.

TENDERS FOR THE CONSTRUCTION /REHABILITATION OF DAMS

TABLE 3 ANALYSIS OF BILL ITEMS (AFTER CORRECTION) COMPARED WITH THE ENGINEER'S ESTIMATE

Bills	Engr's Est	1	2	3	4	5	6
1 Earthworks							
1.i.a							
b							
ii							
iii							
iv							
Total							

NOTE: 1. Numbers 1–6 represent the names of bidders as listed in Table 1; the table can be expanded to accommodate the actual names.

2. The table should be elongated to accommodate all bill items.

See Table C, for example

A summary of recommendations is given in Table D.

TENDERS FOR THE CONSTRUCTION /REHABILITATION OF DAMS

APPENDIX TO TENDER ANALYSIS FORMAT

MINIMUM EQUIPMENT REQUIRED PER DAM OR VALLEY TANK SITE

The construction and rehabilitation of small earth dams is basically an earth-moving operation supported by wetting and compaction to obtain optimum watertightness. The following equipment and quantities are therefore the minimum required to build/rehabilitate a dam within 6 to 8 weeks:

Equipment*	No
- 4 x w drive double cabin p/up	1
- bulldozer	2
- payloader	2
- tippers (4 m^3 or more)	3
- motor grader	1
- excavator	1 (or more)
- sheepfoot compactor	1 motorised or 2 tractor drawn
- water bowser	1 (6,000 lit.)
- de-watering pump	1 (could be on standby)
- fuel storage tank of at least 2,000 l	1

*1. The equipment should be regularly serviced and there should be a competent mechanic on site at all times.

2. A levelling instrument and competent survey technician should be on site throughout the period of construction.

Tables A–D give a summary of a tender analysis and appraisal report, based on the format presented above.

TABLE A BID PRICES (AS READ OUT ON OPENING) – Uganda capital and currency are used here but any country/currency are equally applicable

BIDDER IDENTIFICATION			READ-OUT BID PRICE(S)		MODIFICATIONS OR COMMENTS
NAME	CITY/STATE OR PROVINCE	COUNTRY	CURRENCY(IES) US SH (%)	AMOUNT(S)% US $ (%)	
1. Bidder 1	Kampala	Uganda	-	-	
2. Bidder 2	Kampala	Uganda	3,152,382,860 Ug Shs (100%)	-	
3. Bidder 3	Kampala	Uganda	1,475,330,786.4 Ug Shs (20%)	5,647,199.18 USD (80%)	Exchange rate applicable 1 USD = 1045 Shs. Bidder's prices have been increased by 17% being amount of VAT applicable as stated in the bid.
4. Bidder 4	Kampala	Uganda	489,801.180 Ug Shs (100%)	-	
5. Bidder 5	Kampala	Uganda	742,985,980 Ug Shs (100%)	-	
6. Bidder 6	Kampala	Uganda	878,166,707 Ug Shs (30% of contract price)	USD 2,008,878.087 (70%)	Exchange rate used 1USD = 1020 Shs
7. Bidder 7	Kampala	Uganda	1,220,045,585 Ug Shs	-	
8. Bidder 8	Kampala	Uganda	518,507,550 Ug Shs (100%)	-	
9. Bidder 9	Kampala	Uganda	535,749,900 Ug Shs (100%)	-	
10. Bidder 10	Kampala	Uganda	1,247,549,050 Ug Shs (100%)		

TABLE B   REPORT OF PRELIMINARY EXAMINATION OF BIDS

BIDDER (A)	VERIFICATION (B)	ELIGIBILITY (C)	BID SECURITY (D)	COMPLETENESS OF BID (E)	SUBSTANTIAL RESPONSIVENESS (F)	ACCEPTABLE FOR DETAILED EXAMINATION (G)
Bidder 1[1]	-	-	-	-	-	No
Bidder 2	Yes	Yes	No[2]	Yes	Yes	No
Bidder 3	Yes	Yes	Yes	Yes	Yes	Yes
Bidder 4	No[3]	Yes	No[4]	Yes	Yes	No
Bidder 5	No[5]	Yes	No[6]	Yes	Yes	No
Bidder 6	Yes	Yes	Yes	Yes	Yes	Yes
Bidder 7	Yes	Yes	No[7]	Yes	Yes	No
Bidder 8	Yes	Yes	Yes	Yes	Yes	Yes
Bidder 9	Yes	Yes	Yes	Yes	Yes	Yes
Bidder 10	Yes	Yes	Yes	Yes	Yes	Yes

Notes: 1. Bidder just wrote a letter to inform the secretary of the Central Tender Board of their inability to participate.
2. Presented an unacceptable format of the bid security and whose validity is only 90 days instead of 118 days.
3. Form of bid is not signed.
4. No bid bond security.
5. Form of bid is not signed.
6. No bid security.
7. Bid security format is not acceptable and is valid for only 90 days.

TABLE C   COMPARISON OF CORRECTED BID AMOUNTS WITH ENGINEER'S ESTIMATE

SITE	ENGR'S ESTIMATE	BIDDER 10	BIDDER 9	BIDDER 8	BIDDER 3	BIDDER 6
DISTRICT A						
1. Rwamuranda	150,475,984	160,574,480	-	135,916,550	1,536,935,012	177,167,650
2. Kyera	90,993,375	104,182,430	-	90,741,200	474,163,014	115,635,795
3. Kenwa	156,181,075	188,742,840	-	152,716,300	602,006,916	252,904,465
4. Kishangura	101,187,200	117,800,760	-	-	524,719,958	162,354,115
DISTRICT B						
5. Kigaaga	164,844,460	208,782,771		155,185,155	481,060,737	220,192,687
DISTRICT C						
6. Kyambidde	157,450,425	-	349,061,900	-	571,121,215	211,625,645
7. Rwamakara	72,410,613	-	186,688,000	-	425,076,341	99,000,165
DISTRICT D						
8. Dyangoma	154,097,515	-	-	-	320,930,147	178,498,748
9. Rwemitongole	155,656,958	-	-	-	602,973,451	204,540,655
10. Kasensero	118,498,875	-	-	-	398,394,981	158,215,200
DISTRICT E						
11. Nakakabala (T)	164,775,875	-	-	-	409,207,952	213,741,550
12. Kasejjere	176,350,900	-	-	-	629,340,630	244,223,100
13. Wabikunyu	167,826,725	-	-	-	445,209,026	211,205,500
DISTRICT F						
14. Wabale	242,735,955	325,058,800	-	-	772,750,247	296,242,320
15. Migeera	130,911,605	164,002,520	-	-	449,012,571	174,680,385

TABLE D   SUMMARY OF RECOMMENDATIONS

Contract No	Bidder	Structure Location and Type	Recommended Price SHS 000	Remarks
1	Bidder 10	DISTRICT A 1. Rwamuranda ND 2. Kyera ND 3. Kenwa ND 4. Kishangura DR DISTRICT B 5. Kigaaga ND Total for Contract 1	160,574.48 104,182.43 188,742.84 117,800.76 208,782.771 780,083.281	These are his evaluated bid prices (EBP) and are acceptable.
2	Bidder 10	DISTRICT F 6. Migeera ND	164,002,520	This is his acceptable EBP
3	Bidder 6	DISTRICT F 7. Wabale ND	296,242.320	This is his acceptable EBP
4	Bidder 6	DISTRICT C 8. Kyambidde ND 9. Rwamakara DR Total for Contract 3	211,625.645 99,000.165 310,625.81	These are his acceptable EBP and are acceptable.
5	Bidder 6	DISTRICT D 10 Dyangoma DR 11 Rwemitongole ND 12 Kasensero ND Total for Contract 4	178,498.748 204,540.655 158,215.20 541,254.603	These are his EBP and are acceptable to all.
6	Bidder 6	DISTRICT E		

Contract No	Bidder	Structure Location and Type	Recommended Price SHS 000	Remarks
		13 Nakakabala VT	213,741.55	These are his acceptable EBP
		14 Kasejjere ND	244,223.10	
		15 Wabikunyu VT	211,205.50	
		Total for Contract 5	669,170.15	
	Grand total Contracts 1 – 6		2,761,378.684	About 25% higher than engineer's estimate

KEY: ND = NEW DAM   DR = DAM REHABILITATION   VT = VALLEY TANK

Appendix VIII
SAMPLE PAYMENT CERTIFICATES

1. Blank Formats – Forms 1A – 1C
2. General Summary for Whole Project of more than one site, six in this case Forms 2A, 2B and 2C.
3. Completion Certificate for Rwamuranda Dam – Forms 3A and B.
4. Interim payment Certificate for Kenwa Dam – Forms 4A and B.

Note the explanations given at the bottom of certain sheets.

FORM 1A

CONSTRUCTION/REHABILITATION
OF DAMS/VALLEY TANKS AND RELATED WORKS

CONTRACT NO: ……………………….

INTERIM CERTIFICATE NO: ………………………………..
(E.G. AWEC JF/001)

PERIOD ENDING: …………..

CONTRACTOR:…………………………………………………….

General Summary

NO	DESCRIPTION	AMOUNT (SH. 000)
1.	Total Value of Scheduled Work	
2.	Total Value of Contingencies	
3.	Total Value of Day Works Total Amount of Works Executed	
4.	Less retention (10%)	
5.	Less Advance Recovery	
6.	Less Withholding Tax Cumulative Net Total Less Cumulative on Previous Certificate No ………………. NET AMOUNT PAYABLE ON THIS CERTIFICATE	

Amount in words: …………………………………………….
…………………………………………………………………….

Sgd …………………………..   Sgd…………………………..
      (CONTRACTOR)   (CONSULTANT)

Date …………………….…...   Date ……..……………….

FORM 1B

………………………………. PROJECT
CONSTRUCTION/REHABILITATION OF DAMS/VALLEY TANKS AND RELATED WORKS

CONTRACT NO ……………… INTERIM CERTIFICATE NO ………

SUMMARY SHEET

SITE ………………………………..

BILL ITEM	DESCRIPTION	CONTR. AMOUNT	PREV. CERT.	THIS CERT.	CUM. TOTAL TO DATE	AMOUNT
	TOTAL CARRIED TO GENERAL SUMMARY					

Contractor………………   Consultant……………………

Date……………………..   Date …………………………….

FORM 1C

………………………………. PROJECT

CONSTRUCTION/REHABILITATION OF DAMS/VALLEY TANKS AND RELATED WORKS

CONTRACT NO …………… INTERIM CERTIFICATE NO …………

SUMMARY SHEET

SITE …………………………………..

BILL ITEM	DESCRIPTION	AMOUNT (SH. 000)
	TOTAL CARRIED FORWARD TO GENERAL SUMMARY	

Contractor……………… Consultant……………………

Date…………………….. Date ……………………………

FORM 2A

AGRICULTURAL WATER SUPPLY PROJECT

CONSTRUCTION/REHABILITATION OF DAMS/VALLEY TANKS AND RELATED WORKS

CONTRACT NO: 0050

INTERIM CERTIFICATE NO: AWEC/XXY/00/10

PERIOD ENDING: 30/11/98

CONTRACTOR: Agricultural and Water Engineering Constructions Ltd.

Executive Summary

NO	DESCRIPTION	AMOUNT (SH. '000)
1.	Total Value of Scheduled Work	828,957.992
2.	Total Value of Contingencies	102,734.588 (subject to revision)
3.	Total Value of Day Works	Nil
	Total Amount of Works Executed	931,691.888
	Less Cum on Prev. Cert. AWEC/XXY/001,2,3,4,5,6,7,8,9	887,719.208
		43,927.680
4.	Less retention (10%)	4,392.768
5.	Less Advance Recovery	4,392.768
6.	NET AMOUNT	35,182.644
7.	Less Withholding tax (4%)	1,407.306
NET AMOUNT PAYABLE ON THIS CERTIFICATE		33,775.338

Amount in words: Thirty-three million, seven hundred and seventy-five thousand, three hundred and thirty-eight Shillings

Sgd ……………………………..   Sgd…………………………….
    (CONTRACTOR)   (CONSULTANT)

Date ………………………..…….   Date ……..…………………….

Author's note: (RELATES TO FORM 2A)

1. This executive summary gives the general financial status of the project and the amount of money approved by the consultant (supervisor) on this certificate, i.e. Sh. 33,775.338. Details of the contractor's actual claim and the consultant's approval are shown in form 2C.

2. In order to make pre-payment vetting easy for the employer's auditors, it was agreed in the case of the project from which these examples were taken, that the contractor should prepare neat copies after the consultant's vetting (as shown in this form) – which the consultant would recheck before passing them to the employer for payment. Thus rather than present the with the consultant's cancellations, insertions, questions etc., it is the corrected version (as shown above) that would be signed by contractor and consultant and presented to the employer for payment.

FORM 2B

AGRICULTURAL WATER SUPPLY PROJECT CONSTRUCTION/REHABILITATION OF DAMS/VALLEY TANKS AND RELATED WORKS

CONTRACT NO: 0050

INTERIM CERTIFICATE NO: AWEC/XXY/00/10

PERIOD ENDING: 30/11/98

CONTRACTOR: Agricultural and Water Engineering Constructions Ltd.

SUMMARY OF SCHEDULED WORK

SITE NO	SITE NAME	AMOUNT TO DATE (SH.'000)
1.	Rwamuranda	
2.	Migeera	
3.	Kyera	
4.	Kigaaga	
5.	Kenwa	
6.	Kishangura	
	TOTAL	

Summary of Contingencies

1.	Rwamuranda	
2.	Migeera	
3.	Kyera	
4.	Kigaaga	
5.	Kenwa	
6.	Kishangura	
	TOTAL	

Author's note:

This sheet is designed to help the employer's accountants and auditors to see at a glance the two major claim items – 1 & 2 on executive summary – scheduled work as in bill of quantities; contingencies are works which came up in the course of execution. It should be completed for every certificate.

FORM 2C

AGRICULTURAL WATER SUPPLY PROJECT CONSTRUCTION/REHABILITATION OF DAMS/VALLEY TANKS AND RELATED WORKS

CONTRACT NO: 0050

INTERIM CERTIFICATE NO: AWEC/XXY/00/10

PERIOD ENDING: 30/11/98

CONTRACTOR: Agricultural and Water Engineering Constructions Ltd.

GENERAL SUMMARY

NO	SITE	CONTRACT AMOUNT (SH.'000)	AMT EXEC. UP TO CERT (SH.'000)	WORK EXEC TO DATE (%)	NET AMT PAYABLE (SH.'000)
1.	RWAMURANDA	160,574.480	151,138.816	100	-
2.	MIGEERA	164,002.520	218,526.472	100	-
3.	KYERA	104,182.430	104,115.30	100	810.50
4.	KIGAAGA	208,782.771	197,109.00		9,196.75
5.	KENWA	188,742.840	183,909.70		30,997.45
6.	KISHANGURA	117,800.760	86,416.00		11,414.80
	TOTAL	944,085.801			51,609

	Contractor's Claim	Consultant's Approval
TOTAL AMOUNT EXECUTED	941,215.288	931,691.888
LESS PREVIOUS CERT (1-9)	887,719.208	887,719.208
NET AMOUNT	53,496.08	43,927.680
LESS RETENTION (10%)	5,349.608	4,392.768
SUB-NET	48,146.472	39,575.412
LESS ADVANCE DEDUCTION (10%)	5,349.608	4,392.768
SUB-NET	42,796.864	35,182.644
LESS WITH TAX (4%)	1,711.88	1,407.306
AMT PAYABLE THIS CERT.	41,084.984	33,775.338

Author's note:

This form gives the details of the Executive Summary, i.e. form 2A.

FORM 3A

AGRICULTURAL WATER SUPPLY PROJECT CONSTRUCTION/REHABILITATION OF DAMS/VALLEY TANKS AND RELATED WORKS

CONTRACT NO: 0050

INTERIM CERTIFICATE NO: AWEC/XXY/00/10 (Completion Certificate)

PERIOD ENDING: 30/11/98

CONTRACTOR: Agricultural and Water Engineering Constructions Ltd.

SITE SUMMARY SHEET

SITE: RWAMURANDA

BILL ITEM	DESCRIPTION	UNIT	WORK EXEC. (%)	QTY	RATE (SH. 000)	AMOUNT (SH. 000)
1.01	Clear bush and grub topsoil	ha	100	14.3	480	6,864
1.02	Excavation of foundation trench	m³	100	879.84	4,950	4,355,208
1.04	Borrow, cart, fill & comp. Dam core	m³	100	2,192	7.45	18,698.308
1.05	Ditto for dam wall, using murram	m³	100	4,167	6.0	25,002
1.06	Cut spillway & cart to spoil or reuse	m³	100	11,772	4.9	57,682.800
1.07	Trim slopes to specification	m²	100	3,676	1.0	3,676
1.08	Provide, lay & compact lateritic soil	m³	100	98	4.5	441
1.09	Topsoiling & grassing	m²	100	11,808	1.0	11,808
2.01	Supply & build 150 mm pipe	m	100	120	20	2,400
2.02	Supply & fix cast iron gate valves					
	i. 150 mm diameter	No	100	1	110	110
	ii. 50 mm diameter	No	100	1	55	55

BILL ITEM	DESCRIPTION	UNIT	WORK EXEC. (%)	QTY	RATE (SH. 000)	AMOUNT (SH. 000)
2.03	Supply & fix aluminium strainer	No	100	1	300	300
2.04	Supply & fix PVC reducer tee	No	100	1	110	110
3.01	Provide & fix 50 mm dia. PVC pipe	m	100	120	12	1,440
3.02	Construct watering trough	L/Sum	100	–	–	10,000
3.03	Construct sump	L/Sum	100	–	–	320
4.01	Construct fence as specified	m	100	1,593	4.50	7,168.50
4.02	Fix hard wood gate –			1	8.0	8
	Housing for gate valve	L/Sum	100	–	–	200
	Plumbing in trough –additional	L/Sum	100	–	–	300
	Human tap stand	L/Sum	100	–	–	200
	TOTAL Carried to Gen. Summary		100			151,138.816

Sgd .... Sgd ....

(CONTRACTOR) (CONSULTANT)

Date .... Date ....

Author's note: (relates to FORM 3A)

1. This and the following three forms show the make-up of the contractor's claim in respect of scheduled work and contingencies on two dams given as examples. These dams are used extensively in chapters 5, 6 & 7, hence their choice as examples.

2. This form for Rwamuranda is in fact a final certificate because all works are 100% completed.

FORM 3B

AGRICULTURAL WATER SUPPLY PROJECT
CONSTRUCTION/REHABILITATION OF DAMS/VALLEY TANKS AND RELATED WORKS

CONTRACT NO: 0050

INTERIM CERTIFICATE NO: AWEC/XXY/00/10

PERIOD ENDING: 30/11/98

CONTRACTOR: Agricultural and Water Engineering Constructions Ltd.

SITE SUMMARY OF CONTINGENCIES

SITE: RWAMURANDA

Bill Item	DESCRIPTION	AMOUNT (Sh. 000)
	Excavation of foundation trench – 359.84 m^3 @ Sh. 4,950	1,781.208
	Borrow, cart, fill and compact good clay for dam core – 359.84 @ 7450	2,680.808
	Housing for Gate Valve	200
	Plumbing in troughs	300
	Human tap stand	200
	TOTAL CARRIED TO GENERAL SUMMARY	5,162.016

Sgd …………………………..   Sgd………………………….
      (CONTRACTOR)   (CONSULTANT)

Date ……………………….…..   Date ………...………………….

FORM 4A

AGRICULTURAL WATER SUPPLY PROJECT
CONSTRUCTION/REHABILITATION OF DAMS/VALLEY TANKS
AND RELATED WORKS

CONTRACT NO: 0050

INTERIM CERTIFICATE NO: **AWEC/XXY/00/10**

PERIOD ENDING: 30/11/98

CONTRACTOR: Agricultural and Water Engineering Constructions Ltd.

SITE SUMMARY OF CONTINGENCIES

SITE: KENWA

BILL ITEM	DESCRIPTION	UNIT	WORK EXEC. (%)	QTY	RATE (SH, '000)	AMOUNT (SH.'000)
1.01	Clear bush and grub topsoil	ha	100	1.5	500	750
1.02	Clear bush for access road	ha	100	7.3	480	3,504
1.03	Grub top soil	ha	100	1.0	480	480
1.04	Ditto for reservoir	ha	100	11	480	5,280
1.05	Excavation of foundation trench	m³	100	684	4.95	3,385.80
1.06	Remove old embankment	m³	100 (85)	6,360 (5,460)	4.95	31,482(26,760)
1.07	Borrow, cart, fill & comp. Dam core	m³	100	2,458	7.45	18,312.10
1.08	Ditto for dam wall, using murram	m³	100 (90)	7,737 (6,963)	6.0	46,422(26,760)
1.09	Cut spillway & cart to spoil or reuse	m³	100	6,522	4.90	31,957.80
1.10	Trim slopes to specification	m²	100	3,437	1.0	3,437
1.11	Provide, lay & compact lateritic soil	m³	100	115	4.5	517.50
1.12	Topsoiling & grassing	m²	100	13,934	1.0	13,934
2.01	Supply & build 150 mm pipe	m	100	90	20.0	1,800
2.02	Supply & fix cast iron gate valves (i) 150 mm diameter	No	100	1	110	110
	(ii) 50 mm diameter	No	100	1	55	55
2.03	Supply & fix aluminium strainer	No	100	1	300	300
2.04	Supply & fix PVC reducer tee	No	100	1	110	110
3.01	Provide & fix 50 mm dia. Pvc pipe	m	100	100	12	1,200
3.02	Construct watering trough	L/Sum	100	-	10,000	10,000
3.03	Construct sump	L/Sum	100	-	320	320
4.01	Construct fence as specified	m	100	2,345 (1,407)	4.5	10,552.5
4.02	Fix hard wood gate –	No	100	1	8.0	8
	TOTAL Carried to Gen. Summary		100			183,909.70

Sgd……………………………… Sgd……………………………………
      (CONTRACTOR)   (CONSULTANT)

Date……………………….…... Date……..…………………………..

Author's note:

Although the Contractor claimed 100%, the Consultant approved only 85% in line with actual work done.

FORM 4B

AGRICULTURAL WATER SUPPLY PROJECT
CONSTRUCTION/REHABILITATION OF DAMS/VALLEY TANKS AND RELATED WORKS

CONTRACT NO: 0050

INTERIM CERTIFICATE NO: **AWEC/XXY/00/10**

PERIOD ENDING: 30/11/98

CONTRACTOR: Agricultural and Water Engineering Constructions Ltd.

SITE SUMMARY OF CONTINGENCIES

SITE: RWAMURANDA

BILL ITEM	DESCRIPTION	AMOUNT (SH. 000)
1.08	Cut Spillway and cart to spoil or reuse – 2,517 m^3 @ 4,900	12,333.300
	TOTAL CARRIED TO GENERAL SUMMARY	12,333.300

Sgd ……………………………….. Sgd …………………………….
    (CONTRACTOR)   (CONSULTANT)

Date ……………………………..….. Date ……..…………………

www.ingramcontent.com/pod-product-compliance
Ingram Content Group UK Ltd.
Pitfield, Milton Keynes, MK11 3LW, UK
UKHW041952230426
12048UKWH00008B/288